NUREG–1536

I0489265

Standard Review Plan
for Dry Cask Storage Systems

Final Report

Manuscript Completed: January 1997
Date Published: January 1997

Spent Fuel Project Office
Office of Nuclear Material Safety and Safeguards
U.S. Nuclear Regulatory Commission
Washington, DC 20555–0001

ABSTRACT

The Standard Review Plan (SRP) For Dry Cask Storage Systems provides guidance to the Nuclear Regulatory Commission staff in the Spent Fuel Project Office for performing safety reviews of dry cask storage systems. The SRP is intended to ensure the quality and uniformity of the staff reviews, present a basis for the review scope, and clarification of the regulatory requirements.

Part 72, Subpart B generally specifies the information needed in a license application for the independent storage of spent nuclear fuel and high level radioactive waste. Regulatory Guide 3.61 "Standard Format and Content for a Topical Safety Analysis Report for a Spent Fuel Dry Storage Cask" contains an outline of the specific information required by the staff. The SRP is divided into 14 sections which reflect the standard application format. Regulatory requirements, staff positions, industry codes and standards, acceptance criteria, and other information are discussed.

Comments, errors or omissions, and suggestions for improvement should be sent to the Director, Spent Fuel Project Office, U.S. Nuclear Regulatory Commission, Washington, DC 20555-0001.

ABSTRACT

TABLE OF CONTENTS

DEFINITIONS

Accident-Level. Term used to include both design-basis accidents and design-basis natural phenomenon events and conditions. [See also "Design-Basis _____."] Resistance, response limit, and functional capability requirements apply for conditions and events that exceed "off-normal" or "Design Event II," as described in ANSI/ANS 57.9.

Basic, or fundamental, safety criteria. The following minimal functions for nuclear safety in the design of an ISFSI or MRS facility:

- Maintain subcriticality.
- Prevent release of radioactive material above acceptable amounts.
- Ensure that radiation rates and doses do not exceed acceptable levels.
- Maintain retrievability of the stored radioactive materials throughout the life of the DCSS.

Benchmarking. Validation of the accuracy of a computer code by comparison of the calculated results with those of relevant experiments.

Bias. ANSI/ANS-8.1 defines bias as "a measure of systematic disagreement between the results calculated by a method and experimental data. The uncertainty in the bias is a measure of both the precision of the calculations and the accuracy of he experimental data." See NUREG/CR-6361 for further discussion of bias. Bias defined as the average of the differences between results and measurements is acceptable, provided that one adequately considers the variation in the differences.

Code. Used generically to refer to national or "consensus" codes, standards, and specifications, or specifically to refer to the ASME Boiler and Pressure Vessel Code.

CDE. Committed Dose Equivalent, defined as the total radiation dose equivalent to the body (or specified part of the body) that will be accumulated over 50 years following an intake of radioactive material.

Confinement The spent fuel cladding must be protected during storage against degradation that leads to gross ruptures or the fuel must be otherwise confined such that degradation of the fuel during storage will not pose operational safety problems with respect to its removal from storage.

Containment The assembly of components of the packaging intended to retain the radioactive material during storage.

Confirmatory Calculations. Calculations made by the reviewer to determine whether the package design and specifications meet the regulations. These calculations do not replace the design calculations and are not intended to endorse the applicant's calculations.

Construction. The assembly, fabrication, or putting together of standard parts or components to form structures of systems of a DCSS

Controlled Area. That area immediately surrounding an ISFSI or MRS for which the licensee exercises authority over use and within which ISFSI operations are performed. See 10 CFR 72.3.

Design-Basis _____. The extreme level of an event or condition for which there is a specified resistance, limit of response, and requirement for a given of continuing capability. (Compares with "Design Events III and IV" as described in ANSI 57.9.)

Design Event (I, II, III, or IV). Conditions and events as defined and used for an ISFSI in ANSI/ANS 57.9 (also applicable to an MRS).

Exclusion Area. [Applies to sites with a reactor only] "That area surrounding the reactor, in which the reactor licensee has the authority to determine al activities including exclusion or removal of personnel and property from the area." [10 CFR 100, with additional descriptors included at 10 CFR 100.3.]

Gross Cladding Defect. A known or suspected cladding condition that results in the fuel not meeting its design-basis criteria for dry cask storage. The cask shielding, criticality, thermal, and radiological design analyses typically assume that the cladding provides sufficient structural integrity to retain the fuel pellets in the fuel assembly geometry for normal and accident conditions[1]. In addition, both individual fuel rods and fuel assemblies should be intact to preclude fuel handling or operational safety problems during loading and unloading operations. It is the responsibility of the licensee to ensure that fuel placed in dry storage meets the design-basis conditions. This definition is applicable to all phases of dry cask storage (from selection and inspection of the fuel before loading until the fuel is unloaded from the cask or the cask is placed in a permanent repository). Alternative means, such as canning, will be required for dry cask storage of fuel that does not meet design-basis conditions.

Hard Receiving Surface for a horizontal or vertical drop need not be an unyielding surface; rather the receiving surface may be modeled as a reinforced concrete pad on engineered fill.

Important Confinement Features. Term used in ANSI/ANS 57.9, but not acceptable to the NRC. (Per RG 3.60, "important to safety" should be substituted for "important confinement features" in the standard.)

Important to Safety [also "Important to Nuclear Safety"]. A function or condition required to store spent fuel of high-level waste safely. To prevent damage to the spent fuel or the high-level waste container during handling and storage, to provide reasonable assurance that spent fuel or high level radioactive waste can be received, handled, packaged, stored, and retrieved without undue risk to the health and safety of the public.

Independent Calculation Calculations separate from the applicant's. Input data should be taken from primary sources such as the package drawings and manufacturer's specifications. Models should be developed separately by the reviewer. To the extent possible, different techniques, codes, and cross section sets or other derived data sets should be used.

Intact Cladding. Spent fuel cladding that does not have gross cladding defects (see Gross Cladding Defects).

Mixed waste. Waste material that is hazardous because it contains both radioactive material as well as chemical, toxic, incendiary, or other hazards.

MofS. Margin of safety, which may be defined as identical to factor of safety, f.s. = capacity/demand (with minimum acceptable MofS ≥ 1.0), or as a true margin, where MofS = f.s.-1 = (capacity/demand) - 1 (with minimum acceptable MofS ≥ 0.0.

NDE: Nondestructive examination: testing, examination, and/or inspection of a component which does not affect the use of the component. NDE can be broadly divided into three categories: visual, surface, and volumetric examinations. [Additional information may be found in the ASME B&PV Code, Section V, Nondestructive Examination, Appendix A.]

NDE related terms in order of increasing severity:

discontinuity: an interruption in the normal physical structure of a material. Discontinuities may be unintentional, such as those formed inadvertently during the fabrication process, or intentional, such as a drilled hole.

indication: detection of any discontinuity using an NDE method.

flaw: detection of an imperfection or unintentional discontinuity using an NDE method.

defect: a flaw which, due to its size, shape, orientation, location, or other properties, is rejectable to the applicable construction code. Defects may be detrimental to the intended service of a component and the component must be repaired or replaced.

---

[1] The statements of consideration for the standardized NUHOMS spent fuel storage cask FR 65898, comment F.4, provides an example of past staff practice regarding the allowable condition of fuel cladding for loading in a cask. "Licensees and Certificate of Compliance generally require that the fuel have no known or suspected gross cladding breaches to ensure the structural integrity of the fuel. Known or suspected failed fuel assemblies (rods) and fuel with cladding defects greater than pin holes and hairline cracks are not authorized in the Standardized NUHOMS."

Common NDE examination methods include:

LT	leak testing
MT	magnetic particle examination
PT	liquid penetrant examination
RT	radiographic examination
UT	ultrasonic examination
VT	visual examination

destructive
examination: testing, examination, and/or inspection of a component which results in the destruction of the component.

Normal. The maximum level of an event or condition expected to routinely occur. The ISFSI or MRS is expected remain fully functional and to experience no temporary or permanent degradation from normal operations, events, and conditions. (Compares to "Design Event I" of ANSI/ANS 57.9.) Events and conditions that exceed "normal" levels are considered to be, and to have the response allowed for, "off-normal" or "accident-level" events and conditions.

Off-Normal. The maximum level of an event or condition that although not occurring regularly can be expected to occur with moderate frequency and for which there is a corresponding maximum specified resistance, limit of response, or requirement for a given level of continuing capability. (Similar to "Design Event II" of ANSI/ANS 57.9.) ISFSI SSC are expected to experience off-normal events and conditions without permanent deformation or degradation of capability to perform their full function (although operations may be suspended or curtailed during off-normal conditions) over the full license period.

Other radioactive wastes. Components generally associated with the spent fuel, e.g. Control Assemblies (Rods) BWR fuel channels etc.

Quality Group. NRC classification of SSCs by degree of importance to nuclear safety (NUREG-0800, §3.2.2, and Regulatory Guide 1.26) for reactor systems and adapted to use with ISFSI as follows:

- *Quality Group B* is accepted for "construction" [see term in Glossary] of ISFSI and MRS confinement vessels and their integral and contained components.

- *Quality Group C* is accepted for construction of fluid systems that may be connected to a penetration of the confinement barrier and used or located outside an NRC-licensed enclosing structure providing tertiary confinement (e.g., the fuel pool building).

- *Quality Group D* is accepted for construction of fluid systems that may be connected to a penetration of the confinement barrier and used or located outside an NRC-licensed enclosing structure providing tertiary confinement, if analysis shows that the maximum conservatively estimated offsite dose (the analysis procedure identified in RG 1.26, Subsection C.2.d, is acceptable) would not exceed 0.5 rem to the whole body or any equivalent part of the body.

Radwaste. Waste that is hazardous because it contains nuclear materials (may be high- or low-level.

Ready Retrievability. Capability to return the stored radioactive material to a safe condition without the release of radioactive materials to the environment or radiation exposures in excess of the limits defined by 10 CFR 20 [10 CFR 72.122(h)(5)]. ISFSI and MRS storage systems must be designed to allow ready retrieval of the stored spent fuel or high-level waste (MRS only) for compliance with 10 CFR 72.122(l).

Restricted Area. "Any area to which the licensee controls access to protect individuals from exposure to radiation and radioactive materials." [10 CFR 20]

Safety Analysis Report. In the context of the FSRP, the report submitted by the license applicant in compliance with 10 CFR 72, Subpart B or I. The fundamental contents of the report are described at 10 CFR 72.24. Guidance regarding the content of the report is provided by Reg. Guides 3.48, 3.61 and 3.62. For the staff review, the SAR is considered to constitute the actual SAR submitted with the application, along with supplemental data submitted with the application and supplemental data and responses submitted following the application during the NRC staff review and evaluation. The effective SAR is considered by the staff to be that submitted, as amplified and/or modified by the supplemental and later submissions that are docketed.

Safety Evaluation Report. In the context of the FSRP, the report prepared by the NRC staff to present findings and recommendations relating to the acceptability of the applicant's safety analysis and other required submissions. The SER also identifies the bases for those recommendations and the recommended technical specifications ("operating controls and limits" or "conditions of use").

Unrestricted Area. "Any area to which the licensee need not control access in order to protect individuals from exposure to radiation and radioactive materials." [10 CFR 20]

Volume %. The percent of a mole of the material that is present in a volume equal to the standard volume for the material as a gas; the volume occupied by one mole of the material as a gas at standard conditions for gases (760 mm Hg (760 torr) pressure and 0°C temperature).

INTRODUCTION

This standard review plan (SRP) provides guidance for use by staff reviewers from the U.S. Nuclear Regulatory Commission (NRC), Office of Nuclear Material Safety and Safeguards (NMSS), Spent Fuel Project Office, in performing safety reviews of applications for approval of spent fuel dry cask storage systems (DCSS). The principal purposes of the DCSS SRP are to ensure the quality and consistency of staff reviews and to establish a well-defined basis from which to evaluate proposed changes in the scope of reviews.

Other purposes of this SRP are to ensure wide availability of information about regulatory matters, to improve communication, and to help interested persons and the nuclear power industry better understand the staff review process.

The regulations (10 CFR Part 72) that govern the storage of spent nuclear fuel are largely performance based. An example of a performance based regulation can be found in 72.122 Overall requirements:

> (a) Quality Standards. Structures, systems, and components important to safety must be designed, fabricated, erected, and tested to quality standards commensurate with the importance to safety of the function to be performed.

This SRP describes the process and provides the reference documents reviewers need to evaluate what "commensurate with importance to safety" means and to evaluate it constantly with respect to the many different designs for DCSS that may be submitted for approval.

A DCSS may be used to store spent nuclear fuel under either a site-specific or general license to operate an independent spent fuel storage installation (ISFSI). At present, any holder of an active reactor operating license under Title 10, Part 50, of the *U.S. Code of Federal Regulations* (10 CFR Part 50), has the authority to construct and operate an ISFSI under the provisions of the general license. Requirements for construction and pre-operational activities of such an ISFSI are discussed in Subparts K and L of 10 CFR Part 72. The requirements for pursuing a site-specific ISFSI license are discussed in Subparts B and C of 10 CFR Part 72. Regardless of the license type, the NRC staff must review and approve the cask design that will be used in an ISFSI before spent fuel loading begins. This SRP describes the methods used by the NRC staff to conduct such a review.

The DCSS safety review is primarily based on the information provided by an applicant, or cask vendor, in a safety analysis report (SAR). Sections 72.24 and 72.230 of 10 CFR Part 72 require inclusion of an SAR in each application for a license to store spent nuclear fuel or for approval of spent fuel casks. Before submitting an SAR, an applicant should have designed and analyzed the storage cask system in sufficient detail to conclude that it can be properly fabricated and safely operated without endangering the health and safety of the public. The SAR is the principal document in which the applicant provides the information that reviewers need in order to understand the bases for reaching the conclusion that the storage cask is acceptable for use.

Section 72.24 specifies, in general terms, the information to be supplied in an SAR. The specific information required by the staff for evaluation of an application is identified in Regulatory Guide (RG) 3.61, "Standard Format and Content of Topical Safety Analysis Reports for a Spent Fuel Dry Storage Facility." The sections of this SRP are keyed to the standard format defined in RG 3.61. Similar information is also provided in RG 3.62, "Standard Format and Content for the Safety Analyses Report for On-Site Storage of Spent Fuel Storage Casks."

This SRP is written to address a variety of site conditions and cask system designs. Each section presents the complete review procedure and all current acceptance criteria for all pertinent areas of review. However, for any given application, the staff reviewers may select and emphasize the particular aspects of each SRP section that are appropriate for a given application. In some cases, a cask feature may be sufficiently similar to that of a previous cask so that a *de novo* review of the feature is not needed. For these and other similar reasons, the staff may not carry out in detail all of the review steps listed in each SRP section in the review of every application. Conversely, the staff may find it necessary to ask additional questions or probe areas in greater depth, in order to adequately review a particular design. Review plans have not been included for SAR sections that consist of background or design data that are included for information or for use in reviewing other SAR sections.

The individual SRP sections address, in detail, the matters that are reviewed, the basis for the review, how the review is accomplished, and the conclusions that are sought. Each SRP section is organized into seven subsections, as follows:

I.   Review Objective

This subsection states the purpose and scope of the review.

II.   Areas of Review

This subsection describes the systems, components, analyses, data, or other information that are reviewed as part of the given SAR section. It also discusses the information needed or coordination expected from reviewers of other SAR sections in order to complete the subject technical review.

III.   Regulatory Requirements

This subsection summarizes the applicable sections of 10 CFR Part 72 pertaining to the given SAR section. This list is not all inclusive (e.g., some parts of the regulations, such as 10 CFR Part 20, are assumed to apply to all sections of the SAR).

IV.   Acceptance Criteria

This subsection addresses the design criteria and in some cases specific analytical methods that NRC staff reviewers have found to be acceptable for meeting the regulatory requirements, specified in 10 CFR Part 72, that apply to the given SAR section.

These acceptance criteria typically set forth the solutions and approaches that staff reviewers have previously determined to be acceptable in dealing with a specific safety problem or design area that is important to safety. These solutions and approaches are discussed in the SRP so that staff reviewers can take uniform and well-understood positions as similar safety issues arise in future cases. Like regulatory guides, these solutions and approaches are acceptable to the staff, but they are not the only possible solutions and approaches. Applicants should recognize that, as in the case of regulatory guides, substantial staff time and effort has gone into developing these acceptance criteria, and a corresponding amount of time and effort may be required to review and accept new or different solutions and approaches. Thus, applicants proposing solutions and approaches to new safety issues or analytical techniques other than those described in the SRP should expect longer review times and more extensive questioning in these areas. An alternative is to propose new methods on a generic basis, apart from a specific license application. Such an alternative proposal could consist of a submittal of a Topical Safety Analysis Report (TSAR). This type of application could form the basis for either a change in the staff interpretation of the regulatory requirements or support a request for rulemaking to change the requirements themselves.

V.   Review Procedures

This subsection discusses how the review is to be accomplished, including the general procedure that reviewers follow to establish reasonable verification that the applicable safety criteria have been met.

VI.   Evaluation Findings

This subsection presents the type of conclusion that is sought for the given review area. For each area, a conclusion of this type is included in the safety evaluation report (SER) in which the staff reviewers publish their findings. The SER also describes which aspects of the review were selected or emphasized; which matters were modified by the applicant, require additional information, will be resolved in the future, or remain unresolved; where the cask's design deviates from the criteria stated in the SRP; and the bases for any deviations from the SRP.

VII. References

This subsection lists the references commonly used in the review process for the given subject area.

The SRP and RG 3.61 are directed toward storage cask systems designed for spent fuel with zircalloy cladding. Staff reviewers may adapt the SRP as needed for use in reviewing other storage designs and spent fuel types.

The SRP results from years of staff experience establishing and using regulatory requirements to review SARs and to evaluate the safety of spent fuel storage system designs. This SRP may be considered a part of the continuing regulatory standards development process and documents current review methods.

The SRP may be revised and updated as the need arises to clarify the content, correct errors, or incorporate modifications approved by the Director of the SFPO. Comments, suggestions for improvement, and notices of errors or omissions will be considered by and should be sent to the Director, Spent Fuel Project Office, Office of Nuclear Material Safety and Safeguards, U.S. Nuclear Regulatory Commission, Washington, DC 20555.

1.0 GENERAL DESCRIPTION

I.   Review Objective

The purpose of reviewing the general description of the cask or dry cask storage system (DCSS) is to ensure that the applicant has provided a non-proprietary description that is adequate to familiarize reviewers and other interested parties with the pertinent features of the system.

II.   Areas of Review

The general description should enable all reviewers, regardless of their specific review assignments, to obtain a basic understanding of the DCSS, its components, and the protections afforded for the health and safety of the public. Regulatory Guide (RG) 3.61[1] provides general guidance regarding information that should be included in the general description. Because much of the information relevant to this initial aspect of the DCSS review is presented in more detail in other chapters of this standard review plan (SRP), this chapter focuses on familiarization with the DCSS and should be consistent with the remaining sections of the safety analysis report (SAR). Specifically, this focus may encompass the following areas of review:

1.   DCSS description and operational features
2.   drawings
3.   DCSS contents
4.   qualifications of the applicant
5.   quality assurance
6.   consideration of 10 CFR Part 71[2] requirements regarding transportation

III.   Regulatory Requirements

1.   General Description and Operational Features

The application must present a general description and discussion of the DCSS, with special attention to design and operating characteristics, unusual or novel design features, and principal safety considerations. [10 CFR Part 72.24(b)][3]

2.   Drawings

Structures, systems, and components (SSCs) important to safety must be described in sufficient detail to enable reviewers to evaluate their effectiveness. [10 CFR Part 72.24(c)(3)]

3.   DCSS Contents

The applicant must provide specifications for the contents expected to be stored in the DCSS (normally spent fuel). These specifications may include, but not be limited to, type of spent fuel (i.e., boiling-water reactor (BWR), pressurized-water reactor (PWR), or both), maximum allowable enrichment of the fuel before any irradiation, burnup (i.e., megawatt-days/metric ton Uranium), minimum acceptable cooling time of the spent fuel before storage in the DCSS (aged at least 1 year), maximum heat designed to be dissipated, maximum spent fuel loading limit, condition of the spent fuel (i.e., intact assembly or consolidated fuel rods), weight and nature of non-spent fuel contents, and inert atmosphere requirements. [10 CFR Part 72.2(a)(1) and 10 CFR Part 72.236(a)]

4.   Qualifications of the Applicant

The application must include the technical qualifications of the applicant to engage in the proposed activities. Qualifications should include training and experience. [10 CFR Part 72.24(j), 10 CFR Part 72.28(a)]

5. Quality Assurance

The safety analysis report (SAR) must include a description of the applicant's quality assurance (QA) program, with reference to implementing procedures. This description must satisfy the requirements of 10 CFR Part 72, Subpart G, and must be applied to DCSS SSC that are important to safety throughout all design, fabrication, construction, testing, operations, modifications and decommissioning activities. These implementing procedures need not be explicitly included in the application. [10 CFR Part 72.24(n)]

6. Consideration of 10 CFR Part 71 Requirements Regarding Transportation

If the DCSS under consideration has previously been reviewed and certified for use as a transportation cask, the application must include a copy of the Certificate of Compliance issued for the DCSS under 10 CFR Part 71, including drawings and other documents referenced in the certificate. [10 CFR 72.230(b)]

IV. Acceptance Criteria

In general, this initial aspect of the DCSS review seeks to ensure that the applicant's general description of the DCSS fulfills the following acceptance criteria:

1. DCSS Description and Operational Features

The applicant should provide a broad overview and a general, non-proprietary description (including illustrations) of the DCSS, clearly identifying the functions of all components and providing a list of those components classified by the applicant as being "important to safety."

2. Drawings

The applicant should provide non-proprietary drawings of the storage system, of sufficient detail, that an interested party can ascertain its major design features and general operations.

3. DCSS Contents

The applicant should characterize the fuel and other radioactive wastes expected to be stored in the DCSS. If the potential exists that the DCSS will be used to store degraded fuel, the SAR should include a discussion of how the sub-criticality and retrievability requirements will be maintained.

4. Discussion of Organizational Roles

The reviewer should ensure that the applicant has clearly identified the roles and responsibilities that the DCSS designer, vendor, and other agents, such as potential licensees, fabricators, and contractors will have in the review process. Verify that the applicant has provided clear evidence demonstrating that they are qualified to engage in the proposed activities. In addition, verify that the applicant has delineated the responsibilities for all those who will be involved in the construction and operation of the DCSS if known. The reviewer should ensure that the applicant has specifically defined activities which they will not perform.

5. Quality Assurance

Verify that the applicant has described the proposed QA program, citing the applicable implementing procedures. This description should satisfy all requirements of 10 CFR Part 72, Subpart G, that apply to the design, fabrication, construction, testing, operation, modification, and decommissioning of the DCSS SSCs that are important to safety.

6. Consideration of 10 CFR Part 71 Requirements Regarding Transportation

If the DCSS under review has previously been evaluated for use as a transportation cask, the submittal should include the Part 71 Certificate of Compliance and associated documents.

V. Review Procedures

1. DCSS Description and Operational Features

Verify that the applicant has provided for this section of the application, a broad overview of the cask or DCSS design that is non-proprietary and may be used as a tool to familiarize interested parties with the features of the proposed DCSS. This description should present the principal characteristics of the DCSS, including its dimensions, weight, and construction materials. In addition, the description should clearly identify all components that the applicant considers important to safety. Features such as confinement, sampling ports, valves, lids, seals, closure mechanisms, shielding, and lifting devices should be identified and described. A clear definition of the primary confinement system is particularly important. Special design features of the DCSS such as non-passive heat removal system, neutron poisons or monitoring instrumentation should also be discussed.

Sketches and diagrams, if presented in this section, should be compared with the detailed drawings presented elsewhere in the SAR. If the application includes proprietary drawings and descriptions, that will remain proprietary upon approval of the license or certificate, the sketches, drawings and diagrams that provide the general description and operational features need not show the proprietary features. This may be achieved by depicting less detail or by illustrating generic components which fulfill the design function that differ from the actual design. However, these representations should show the operational concept and safety-related features in sufficient detail to form an acceptable basis for public review and comment, as necessary for public hearings.

In addition to information on a single DCSS, the application should describe any limitations on the arrangement of DCSS arrays. For some DCSS, this may be minimum spacing between DCSS, maximum density of DCSS in an array, and/or total number of DCSS or amount of spent fuel that may be stored at a single independent spent fuel storage installation (ISFSI). The acceptable limitations should be included among the conditions for use in the SER (see Chapter 12 of this SRP). However, for DCSS systems such as those with metal confinement vessels stored in a concrete vault, information on the configuration of vault compartments and horizontal/vertical arrangement is necessary.

2. Drawings

Drawings are usually presented in Section 1 of the SAR. Although some applicants may submit drawings designated as "proprietary," reviewers should note that any drawings relied on as the technical basis for adding the DCSS design to the "list of approved spent fuel storage DCSS" contained in Subpart K of 10 CFR 72, become part of the public record. Such drawings will not be treated as proprietary and will be made available to the public (10 CFR 2.790(c)). Applicants may submit additional drawings showing greater detail to support their evaluations, and these may be exempted from the public record if they are not relied on by the staff as part of the technical basis for DCSS design approval. Verify that all SSC important to safety that are not detailed in latter sections of the SAR are sufficiently detailed to enable reviewers to evaluate their effectiveness. In addition, information on non-safety items may also be necessary to ensure that they do not impede the safety systems.

The level of detail needed in the drawings is generally assessed by each reviewer during the evaluation of specific sections of the SAR. Particular attention should be devoted to ensuring that dimensions, materials, and other details on the drawings are consistent with those described in both the text of the SAR and those used in supplementary analysis. If size reduction has rendered any information unclear or illegible, reviewers should request that the applicant provide larger or full-size drawings.

Drawings applicable to the SAR review should be identified by number and revision in Section 12 of the SER.

3. DCSS Contents

The application should present a general description of the fuel or other contents proposed for storage in the DCSS. Because a very detailed description of the proposed DCSS contents or spent fuel is typically provided in Section 2, "Principal Design Criteria," the information presented in Section 1, "General Description," is important only to the extent that it permits overall familiarization with the DCSS. Key parameters for spent fuel include the type of fuel (i.e., PWR, BWR, or both), number of fuel assemblies, and condition of the fuel assemblies (intact or consolidated). This section often includes additional characteristics, such as maximum burnup, initial enrichment, heat load, and cooling time, as well as the assembly vendor and configuration (e.g., Westinghouse 17x17), and these characteristics may also be repeated in the principal design criteria. In addition, the cover gas, if any, should be identified.

If the applicant proposes the storage of spent fuel with gross cladding defects or storage of non-fuel core components that do not have an integral confinement boundary, the range of permissible conditions for the stored material must be defined. If the DCSS system is intended to be used to store fuel with gross cladding defects or an integral confinement boundary when placed in the confinement DCSS, the possible range of conditions of the fuel or components should be stated. 10 CFR 72.122(h)(1) requires "canning" or use of other acceptable means for storing fuel with cladding that is not or may not remain intact and for unconsolidated assemblies (without intact cladding). The application, therefore, should address the following basic requirements:

- maintain subcriticality
- prevent unacceptable release of contained radioactive material
- avoid excessive radiation dose rates and doses
- maintain ready retrievability of the contents

If the requested approval is to address the possible use of the DCSS system for storing non-fuel core components, the application should present summary descriptions of those components. Also, if the components are degraded (e.g., the component does not provide adequate confinement under design basis conditions to contain radioactive gas or other dispersable radioactive materials), the application should describe the possible conditions and alternative confinement methods, if any.

4. Qualifications of the Applicant

The application should clearly designate the applicant and the prime agents, consultants, and contractors, if known, for design, fabrication, and testing of the proposed DCSS components. In addition, the application should clearly define the division and assignment of responsibilities among those parties. Although specific subcontractors may not be known at the time the SAR is submitted, the application should clearly identify any activities that the applicant will not perform. In addition, the application should describe the technical qualifications, previous experience and suitability of the all organizations participating in the proposed activities.

5. Quality Assurance

The applicant should describe the proposed QA program, citing all implementing procedures, in a manner that satisfies the 18 criteria defined in 10 CFR Part 72, Subpart G, "Quality Assurance." (The description need only refer to procedures that implement the QA program. These procedures need not be explicitly included in the application.) The QA program should address design, fabrication, construction, testing, operation, modification, and decommissioning activities regarding the DCSS SSCs that are important to safety. The applicant should also discuss the activities that will be performed under the QA program, and how the activities will be controlled to ensure compliance with all of the requirements of Subpart G. These controls should be applied to the various activities using a graded approach, i.e., QA efforts expended for a given activity should be consistent with that activite's system classification and function.

6. Consideration of 10 CFR Part 71 Requirements Regarding Transportation

If the DCSS system under review for storage has previously been evaluated for use as a transportation DCSS, the submittal should include a copy of the Certificate of Compliance issued for the DCSS under 10 CFR Part 71, if applicable. Drawings or other documents referenced in the certificate, should be included with the application or incorporated by reference to the NRC Part 72 docket number. The applicant should include a discussion supporting the proposed use of the DCSS for spent fuel storage for a period of at least 20 years.

Because applications for certification under 10 CFR Parts 71 and 72 are sometimes submitted concurrently, the final (approved) version of such documents may not be available at the time of initial DCSS SAR submission. Nonetheless, applicable documentation of the Part 71 certification, including questions and responses from the related review, is generally provided to the Part 72 review team, as appropriate. Substantial coordination of the Part 71 and Part 72 reviews is necessary to ensure consistency and avoid duplication of effort. The applicant should have a process for promptly informing each of the review teams about DCSS system design changes precipitated by any concurrent safety reviews. Provisions for communicating these changes should be addressed by and discussed with the applicant.

VI. Evaluation Findings

Review the 10 CFR Part 72 acceptance criteria and provide a summary statement for each. These statements should be similar to the following model:

A general description and discussion of the DCSS is presented in Section(s) _____ of the SAR, with special attention to design and operating characteristics, unusual or novel design features, and principal safety considerations.

Drawings for structures, systems, and components (SSCs) important to safety are presented in Section _____ of the SAR. A listing of those drawings that were relied upon as a basis for approval appears in Section 12 of the Safety Evaluation Report (SER).

Specifications for the spent fuel to be stored in the DCSS are provided in SAR Section _____. Additional details concerning these specifications are presented in Chapter 2 of both the SAR and SER.

The technical qualifications of the applicant to engage in the proposed activities are identified in Section _____ of the SAR.

The quality assurance program, and implementing procedures, are described in Section 13 of the SAR.

The [DCSS system designation] [has been/is/is not being] certified under 10 CFR Part 71 for use in transportation. A copy of the SAR and Certificate of Compliance issued under 10 CFR Part 71 are on file with the NRC under Docket No. _____ [if applicable].

The staff concludes that the information presented in this section of the SAR satisfies the requirements for the general description under 10 CFR Part 72. This finding is reached on the basis of a review that considered the regulation itself, Regulatory Guide 3.61, and accepted practices.

VII. References

1.   U.S. Nuclear Regulatory Commission, "Standard Format and Content for a Topical Safety Analysis Report for a Spent Fuel Dry Storage Cask," Regulatory Guide 3.61, February 1989.

2.   *U.S. Code of Federal Regulations*, "Packaging and Transportation of Radioactive Material," Part 71, Title 10, "Energy."

3.   *U.S. Code of Federal Regulations*, "Licensing Requirements for the Independent Storage of Spent Nuclear Fuel and High-level Radioactive Waste," Part 72, Title 10, "Energy."

2.0 PRINCIPAL DESIGN CRITERIA

I. Review Objective

The purpose of evaluating the principal design criteria related to structures, systems, and components (SSC) important to safety is to ensure that they comply with the relevant general criteria established in 10 CFR Part 72[1], further guidance can be found in NUREG/CR-6407[2] "Classification of Transportation Packaging and Dry Spent Fuel Storage System Components According to Importance to Safety." Material provided in this chapter will form the basis for accepting the safety analysis report (SAR) for staff review.

The applicant should present details of the principal design criteria in either Section 2 or defer the details to the associated sections of the SAR. If the applicant chooses deferral, a general reference to these criteria must be presented. Regulatory Guide (RG) 3.61[3] provides general guidance concerning information that should be included in the principal design criteria for a dry cask storage system (DCSS). In general, these criteria include specifications regarding the fuel or other material to be stored in the DCSS, as well as the external conditions that may exist in the casks operating environment during normal and off-normal operations, accident conditions, and natural phenomena events. A detailed evaluation of how the DCSS design meets the principal design criteria should be presented in Sections 3 through 14 of the safety evaluation report (SER).

II. Areas of Review

The following areas of review have been adopted by the NRC staff, and include those areas noted in RG 3.61:

1. structures, systems, and components important to safety

2. design bases for structures, systems, and components important to safety
    a. spent fuel specifications
    b. external conditions

3. design criteria for safety protection systems

    a. general
    b. structural
    c. thermal
    d. shielding/confinement/radiation protection
    e. criticality
    f. operating procedures
    g. acceptance tests and maintenance
    h. decommissioning
    i. material compatibility[4]

III. Regulatory Requirements

1. Structures, Systems, and Components Important to Safety

The applicant must identify all SSC that are important to safety, and describe the relationships of non-important to safety SSC on overall DCSS performance. [10 CFR 72.24(c)(3) and 72.44(d)]

The applicant must specify the design bases and criteria all SSC that are important to safety. [10 CFR 72.24(c)(1), 72.24(c)(2), 72.120(a), and 72.236(b)]

2. Design Bases for Structures, Systems, and Components Important to Safety

a. Spent Fuel Specifications

The applicant must provide the range of specifications for the spent fuel to be stored in the DCSS. These specifications should include, but are not to be limited to: the type of spent fuel (i.e., boiling-water reactor (BWR), pressurized-water reactor (PWR), or both); content, weight, dimensions and configurations of the fuel; maximum allowable enrichment of the fuel before any irradiation; maximum fuel burnup (i.e., megawatt-days/mtu); minimum acceptable cooling time of the spent fuel before storage in the DCSS (aged at least 1 year); maximum heat load to be dissipated; maximum spent fuel elements to be loaded; spent fuel condition (i.e., intact assembly or consolidated fuel rods); and any inerting atmosphere requirements. [10 CFR 72.2(a)(1) and 72.236(a)]

b. External Conditions

The design bases for SSC important to safety must reflect an appropriate consideration of environmental conditions associated with normal operations, as well as design considerations for both normal and accident conditions and the effects of natural phenomena events. [10 CFR 72.122(b)]

3. Design Criteria for Safety Protection Systems

a. General

The DCSS must be designed to safely store the spent fuel for a minimum of 20 years and to permit maintenance as required. [10 CFR 72.236(g)]

SSC important to safety must be designed, fabricated, erected, and tested to quality standards commensurate with the importance to safety of the function to be performed. [10 CFR 72.122(a)]

The applicant must identify all codes and standards applicable to the SSC. [10 CFR 72.24(c)(4)]

b. Structural

SSC that are important to safety must be designed to accommodate the combined loads of normal operations, accidents, and natural phenomena events with an adequate margin of safety. [10 CFR 72.24(c)(3), 72.122(b), and 72.122(c)]

The design-basis earthquake must be equivalent to or exceed the safe shutdown earthquake of a nuclear plant at sites evaluated under 10 CFR Part 100[5]. [10 CFR 72.102(f)]

The DCSS must maintain confinement of radioactive material within the limits of 10 CFR Part 72 and Part 20, under normal, off-normal, and credible accident conditions. [10 CFR 72.236(l)]

The DCSS must be designed and fabricated so that the spent fuel is maintained in a subcritical condition all under all credible normal, off-normal, and accident conditions. [10 CFR 72.124(a) and 72.236(c)]

The spent fuel cladding must be protected during storage against degradation that leads to gross ruptures, or the fuel must be otherwise confined such that degradation of the fuel during storage will not pose operational safety problems with respect to its removal from storage. [10 CFR 72.122(h)(1)]

Storage systems must be designed to allow ready retrieval of spent fuel waste for further processing or disposal. [10 CFR 72.122(l)]

c. Thermal

Each spent fuel storage or handling system must be designed with a heat removal capability having testability and reliability consistent with its importance to safety. [10 CFR 72.128(a)(4)]

The DCSS must be designed to provide adequate heat removal capacity without active cooling systems. [10 CFR 72.236(f)]

d. Shielding/Confinement/Radiation Protection

The proposed DCSS design must provide radiation shielding and confinement features that are sufficient to meet the requirements of 10 CFR 72.104 and 72.106. [10 CFR 72.126(a), 72.128(a)(2), 72.128(a)(3), and 72.236(d)]

During normal operations and other anticipated occurrences, the annual dose equivalent to any real individual who is located beyond the controlled area must not exceed 25 mrem to the whole body, 75 mrem to the thyroid, and 25 mrem to any other organ as a result of exposure to (1) planned discharges to the general environment of radioactive materials except radon and its decay products, (2) direct radiation from operations of the ISFSI or monitored retrievable storage (MRS), and (3) any other radiation from uranium fuel cycle operations within the region. [10 CFR 72.24(d), 72.104(a), and 72.236(d)]

Any individual located at or beyond the nearest boundary of the controlled area shall not receive a dose greater than 5 rem to the whole body or any organ from any design-basis accident. The minimum distance from the spent fuel handling and storage facilities to the nearest boundary of the controlled area shall be 100 meters. [10 CFR 72.24(d), 72.24(m), 72.106(b), and 36(d)]

The DCSS must be designed to provide redundant sealing of confinement systems. [10 CFR 72.236(e)]

Storage confinement systems must have the capability for continuous monitoring in a manner such that the licensee will be able to determine when corrective action needs to be taken to maintain safe storage conditions. [10 CFR 72.122(h)(4) and 72.128(a)(1)]

The DCSS design must include inspections, instrumentation and/or control (I&C) systems to monitor the SSC that are important to safety over anticipated ranges for normal and off-normal operation. In addition, the applicant must identify those control systems that must remain operational under accident conditions. [10 CFR 72.122(i)]

e. Criticality

Spent fuel transfer and storage systems must be designed to remain subcritical under all credible conditions. [10 CFR 72.124(a) and 72.236(c)]

When practicable, the DCSS must be designed on the basis of favorable geometry, permanently fixed neutron-absorbing materials (poisons), or both. Where solid neutron-absorbing materials are used, the design shall allow for positive means to verify their continued efficacy. [10 CFR 72.124(b)]

f. Operating Procedures

The DCSS must be compatible with wet or dry spent fuel loading and unloading procedures. [10 CFR 72.236(h)]

Storage systems must be designed to allow ready retrieval of spent fuel for further processing or disposal. [10 CFR 72.122(l)]

The DCSS must be designed to minimize the quantity of radioactive waste generated. [10 CFR 72.24(f) and 72.128(a)(5)]

The applicant must describe equipment and processes proposed to maintain control of radioactive effluents. [10 CFR 72.24(l)(2)]

To the extent practicable, the DCSS must be designed to facilitate decontamination. [10 CFR 72.236(l)]

The applicant must establish operational restrictions to meet the limits defined in 10 CFR Part 20 and to ensure that radioactive materials in effluents and direct radiation levels associated with ISFSI operations will remain as low as is reasonably achievable (ALARA). [10 CFR 72.24(e) and 72.104(b)]

g. Acceptance Tests and Maintenance

The DCSS design must permit testing and maintenance as required. [10 CFR 72.236(g)]

SSC that are important to safety must be designed, fabricated, erected, tested, and maintained to quality standards commensurate with the importance to safety of the function to be performed. [10 CFR 72.24(c), 72.122(a), 72.122(f), and 72.128(a)(1)]

h. Decommissioning

The DCSS must be compatible with wet or dry unloading facilities. [10 CFR 72.236(h)]

The DCSS must be designed for decommissioning. Provisions must be made to facilitate decontamination of structures and equipment and to minimize the quantity of radioactive wastes, contaminated equipment, and contaminated materials at the time the ISFSI is permanently decommissioned. [10 CFR 72.24(f), 72.130, and 72.236(I)]

The applicant must provide information concerning the proposed practices and procedures for decontaminating the site and facilities and for disposing of residual radioactive materials after all spent fuel has been removed. Such information must provide reasonable assurance that decontamination and decommissioning will adequately protect the health and safety of the public. [10 CFR 72.24(q) and 72.30(a)]

IV. Acceptance Criteria

The reviewer should verify that the applicant has provided either general or summary discussions of the SSC's design features, and both operational and accident conditions in a sufficiently clear manner that the applicant demonstrates a clear and defensible case that they have met the design criteria. In evaluating the principal design criteria related to DCSS SSC that are important to safety, reviewers should seek to ensure that the given design fulfills the following acceptance criteria:

1. Structures, Systems, and Components Important to Safety

The applicant should discuss the general configuration of the DCSS, and should provide an overview of specific components and their intended functions. In addition, the applicant should identify those components deemed to be important to safety, and should address the safety functions of those components in terms of how they meet the general design criteria and regulatory requirements discussed above. Additional information concerning specific functional requirements for individual DCSS components are addressed in the subsequent chapters of this SRP.

2. Design Bases for Structures, Systems, and Components Important to Safety

Detailed descriptions of each of the items listed below are generally found in specific sections of the SAR; however, a brief description of these areas, including a summary of the analytical techniques used in the design process, should also be captured in Section 2 of the SAR. This description gives reviewers a perspective on how specific DCSS components interact to meet the regulatory requirements of 10 CFR Part 72. This discussion should be non-proprietary since it may be used to familiarize interested persons with the design features and bounding conditions of operation of a given DCSS.

a. Spent Fuel Specifications

The applicant should define the range and types of spent fuel or other radioactive materials that the DCSS is designed to store. In addition, these specifications should include, but are not to be limited to, the type of spent fuel (i.e., boiling-water reactor (BWR), pressurized-water reactor (PWR), or both), weights of the stored materials, dimensions & configurations of the fuel, maximum allowable enrichment of the fuel before any irradiation, burnup (i.e., megawatt-days/mtu), minimum acceptable cooling time of the spent fuel before storage in the DCSS (aged at least 1 year), maximum heat designed to be dissipated, maximum number of spent fuel elements, condition of the spent fuel (i.e., intact assembly or consolidated fuel rods), inerting atmosphere requirements, and the maximum amount of fuel permitted for storage in the DCSS. For DCSSs that will be used to store radioactive materials other than spent fuel, that is, activated components associated with a spent fuel assembly (e.g., control rods, BWR fuel channels), the applicant should specify the types and amounts of radionuclides, heat generation and the relevant source strengths and radiation energy spectra permitted for storage in the DCSS.

b. External Conditions

The SAR should define the bounding conditions under which the DCSS is expected to operate. Such conditions include both normal and off-normal environmental conditions, as well as accident conditions. In addition, the applicant should consider the effects of natural events, such as tornadoes, earthquakes, floods, and lightning strikes. The effects of such events are addressed in individual chapters of the SRP (e.g., the effects of an earthquake on the DCSS structural components are addressed in Chapter 3, "Structural Analysis").

3. Design Criteria for Safety Protection Systems

a. General

The SAR should define an expected lifetime for the cask design. The staff has accepted a minimum of 20 years as consistent with the licensing period. The applicant should also briefly describe the proposed quality assurance (QA) program, and applicable industry codes and standards, that will be applied to the design, fabrication, construction, and operation of the DCSS.

In establishing normal and off-normal conditions applicable to the design criteria for DCSS designs, applicants should account for actual facility operating conditions. Design considerations should therefore reflect normal operational ranges, including any seasonal variations or effects.

b. Structural

The SAR should define how the DCSS structural components are designed to accommodate combined normal, off-normal, and accident loads, while protecting the DCSS contents from significant structural degradation, criticality, and loss of confinement, while preserving retrievability. This discussion is generally a summary of the analytical techniques and calculational results from the detailed analysis discussed in SAR Section 3 and should be presented in a non-proprietary forum.

c. Thermal

The applicant should provide a general discussion of the proposed heat removal mechanisms, including the reliability and verifiability of such mechanisms and any associated limitations. All heat removal mechanisms should be passive and independent of intervening actions under normal and off-normal conditions.

d. Shielding/Confinement/Radiation Protection

The applicant should describe those features of the cask that protect occupational workers and members of the public against direct radiation dosages and releases of radioactive material, and minimize the dose after any off-normal or accident conditions.

e. Criticality

The SAR should address the mechanisms and design features that enable the DCSS to maintain spent fuel in a subcritical condition under normal, off-normal, and accident conditions.

f. Operating Procedures

The applicant should provide potential licensees with guidance regarding the content of normal, off-normal, and accident response procedures. Cautions regarding both loading, unloading, and other important procedures should be mentioned here. Applicants may choose to provide model procedures to be used as an aid for preparing detailed site-specific procedures.

g. Acceptance Tests and Maintenance

The applicant should identify the general commitments and industry codes and standards used to derive acceptance, maintenance, and periodic surveillance tests used to verify the capability of DCSS components to perform their designated functions. In addition, the applicant should discuss the methods used to assess the need for such tests with regard to specific components.

h. Decommissioning

Casks should be designed for ease of decontamination and eventual decommissioning. The applicant should describe the features of the design that support these two activities.

V. Review Procedures

All members of the review team should review Section 2 of the SAR. Although RG 3.61 defines the standard format and content of an SAR, it does not address the different levels of detail expected in introducing component design criteria in SAR Section 2 and as compared with latter sections of the SAR. Consequently, reviewers for each section of the SAR should consider Section 2 in combination with additional details presented later in the SAR. In this SRP, evaluation of design criteria applicable to each of the relevant chapters of the SAR are discussed in detail in those chapters.

Inclusion of a separate section for design criteria in both the SAR and SER supports the staff's procedure of deliberately reviewing these criteria for acceptability apart from the proposed design and infrastructure of the system. This approach forms a "two-step" review process in which the acceptability of the detailed design criteria is separately stated. In-depth evaluation to assess satisfaction of these or other criteria is addressed in other sections of this SRP.

Although the design criteria presented in the SAR may be acceptable to the staff, the actual design may not meet either these criteria or the applicable regulatory requirements. It is also possible that the design criteria themselves, as presented in the SAR, may be unacceptable for application to a given DCSS design. As a result, the design may be unacceptable in that it does not meet the regulatory requirements, or the design may satisfy alternative criteria that are not described in the SAR, but are acceptable to the NRC staff. Reviewers should bring any of these situations to the immediate attention of NRC management.

1. Structures, Systems, and Components Important to Safety

Verify that the applicant has clearly identified all SSC important to safety (as defined by 10 CFR Part 72.3) and documented the rationale for this designation. Such information may be provided in tabular form. Review the general DCSS description presented in SAR Section 1. Ensure that the applicant has provided adequate justification for excluded SSC.

Pay particular attention to instrumentation and other equipment (e.g., lifting devices and transport vehicles). In general, the NRC staff accepts that monitoring systems need not be classified as being important to safety. For example, a failure in the functioning of the pressure monitoring system does not directly result in a release of radionuclides. Additional justification for not considering such systems as being important to safety may be presented in later sections of the SAR and summarized in Section 2.

SSC designated as being important to safety should be included or referenced in the discussion of Design Features within the Technical Specifications provided in SAR Section 12.

2. Design Bases for Structures, Systems, and Components Important to Safety

Verify that the applicant's design bases for DCSS approval accurately identify the range of spent fuel configurations and characteristics, the enveloping conditions of use, and identify the bounding site characteristics. These determine the bounds within which an ISFSI owner may use the SAR, rather than providing additional proof regarding suitability of the covered topics.

a. Spent Fuel Specifications

Review the detailed specifications for the spent fuel to be stored in the DCSS as they are presented in SAR Section 2, and ensure that they are consistent with those discussed in Section 1.The description of the range of spent fuel to be stored should include the type (PWR, BWR, or both), configuration (e.g., 17x17, 15x15, or 8x8), fuel vendor, number of assemblies per cask, enrichment, burnup, minimum cooling time, decay heat generation rate, type of cladding, physical dimensions, total weight per assembly, and uranium weight per assembly. In addition, if control assemblies will be stored with the fuel, ensure that combined weight, dimensions, heat load, and other appropriate information (e.g., number per cask) are specified.

Examine any limitations regarding the condition of the spent fuel. If damage that could be classified as a "Gross Cladding Defect" is allowed, the effects of such damage should be assessed in later sections of the SAR. If damaged rods have been removed from a fuel assembly, determine whether a need exists to replace them with dummy rods before loading into the cask. Note, the presence of an additional moderator will need to be addressed in the criticality analysis in SAR Section 6.

The release of fill and fission product gases from failed fuel rods increases the pressure in the cask cavity, as well as increasing the potential source-term in the event of confinement failure. Consequently, the applicant should provide information regarding the fill/fission product gas present in the fuel as well as the free volume in the cask cavity to enable reviewers to evaluate the pressure in the cask cavity resulting from cladding failure during storage. For the purpose of calculating internal cask pressures, the NRC staff has accepted the following bounding assumptions regarding the minimum percentages of fuel rods to have failed (and released their gases):

- 1% for normal conditions
- 10% for off-normal conditions
- 100% for design-basis (accident and extreme natural phenomena) conditions

Pay particular attention to the specification of burnup, cooling time, and decay heat generation rate. These parameters are generally not independent, and the manner in which they are specified and combined can significantly affect the maximum allowed cladding temperature, as discussed in Chapter 4 of the SRP.

Note the specification of enrichment limits. As discussed in Chapter 5 of the SRP, the criticality evaluation is based on the highest enrichment (for a given fuel assembly), while the shielding source term, especially for neutrons, should be based on the lowest enrichment (for a given burnup).

The SAR will typically list various fuel assemblies that can be stored in the DCSS. In general, no one type of fuel assembly will be bounding for all analyses. Ensure that the applicant has justified which specifications are bounding for each of the evaluations presented in subsequent sections of the SAR. Specifications used in these analyses should also be clearly identified or referenced in Section 12 of both the SAR and SER.

If the applicant requests permission for the storage of non-fuel core components in the cask, review the relevant detailed specifications, conditions, and constraints presented in the SAR. These specifications should be at least as detailed as the applicable information presented for the fuel designs, to provide the reviewer with a basis for a determination that the relevant safety functions of the DCSS will be maintained.

b. External Conditions

The SAR should identify those external conditions that significantly affect, or could potentially affect, the performance of the DCSS. These design-basis conditions will generally restrict either the sites at which the DCSS can be used for spent fuel storage or the manner in which the DCSS can be handled. For example, by selecting the design-basis earthquake (DBE), the SAR limits the use of the cask being reviewed to sites for which the safe shutdown earthquake (SSE) does not exceed the DCSS system DBE. By establishing a design-basis drop, the SAR defines the maximum height to which a cask can be lifted without additional safety analysis or design changes (e.g. impact limiters) by the licensee. Reviewers should note that movement of cask system components within a reactor building may not meet the NRC's criteria for movement of heavy loads within the reactor building[6]. As such, if a potential user (licensee) has been identified, coordination with the appropriate project manager or technical lead from the NRC's Office of Nuclear Reactor Regulation (NRR) should occur during the early stages of DCSS design review.

At a minimum the NRC staff has generally addressed the conditions discussed below; however, other conditions may be relevant depending on specific details of the DCSS design. Reviewers should pay particular attention to special design features and how these might be affected both by other external conditions and other DCSS components.

"Normal" conditions (including conditions involving handling and transfer) and the extreme ranges of normal conditions are presumed to exist during design-basis accidents or design-basis natural phenomena, with the exception of irrational or readily avoidable combinations. For example, an earthquake or tornado may occur at any time and in combination with any "normal" condition. By contrast, it can be presumed that transfer, loading, and unloading operations would not be conducted during a flood.

"Off-normal" conditions and events are presumed to occur in combination with normal conditions that are not mutually exclusive. Nonetheless, it is not required that the SAR analyze or the system be designed for the simultaneous occurrence of independent off-normal conditions or events, design-basis accidents, or design-basis natural phenomena.

Conditions involving a "latent" equipment or instrument failure or malfunction (that is, one that occurs and remains undetected) should be presumed to exist concurrently with other off-normal or design-basis conditions and events.Typical latent malfunctions include a misreading instrument that is not detected as part of routine procedures; an undetected ventilation blockage; or undetected damage from an earlier design-basis event or condition if no provisions exist for detection, recovery, or remediation of such conditions.

For normal , off-normal and accident conditions, reviewers should verify that the applicant has defined appropriate operating and accident scenarios. For these scenarios the applicant should include in the SAR a comprehensive evaluation of the effects of such scenarios on the SSC important to safety. Applicant's evaluations should demonstrate that the requirements of 10 CFR Part 72 .106 as well as 10 CFR Part 20 have been met.

If appropriate, the following design bases should be included as operating controls and limits in Section 12 of both the SAR and SER:

(1) Normal Conditions

For a given spent fuel specification, the primary external conditions that affect DCSS performance are, the ambient temperatures, insolence, and the operational environment experienced by the DCSS.

The NRC accepts as the maximum and minimum "normal" temperatures the highest and lowest ambient temperatures recorded in each year, averaged over the years of record. For the SAR, the applicant may select any design-basis temperatures as long as the restrictions they impose are acceptable to both the applicant and the NRC. If the cask is also designed for transportation, the temperature requirements of 10 CFR Part 71[7] could determine the design basis temperatures for storage.

For storage casks, the NRC staff accepts a treatment of insolence similar to that prescribed in 10 CFR Part 71.71 for transportation casks. If the applicant selects another design approach, it must be justified in the SAR.

The operational environment experienced by the DCSS under normal conditions includes the manner in which the cask is loaded, unloaded, and lifted. Occupational dose rates will in part, depend on whether the cask is sealed in a wet or a dry environment. Fuel cladding temperatures may also be affected. The manner in which the cask is lifted will determine the load on the trunnions and/or lifting yoke. The orientation of the cask (vertical or horizontal) and its height above ground during transport to the ISFSI will establish initial conditions for the drop accidents discussed below.

(2) Off-Normal Conditions

SARs generally address several off-normal conditions. These should include variations in temperatures beyond normal, failure of 10 percent of the fuel rods combined with off-normal temperatures, failure of one of the confinement boundaries, partial blockage of air vents, human error, out-of-tolerance equipment performance, equipment failure, and instrumentation failure or faulty calibration.

(3) Accident Conditions

The staff has generally considered that the following accidents should be evaluated in the SAR. Because of the NRC's defense-in-depth approach, each should be evaluated regardless of whether it is highly unlikely or highly improbable. These do not constitute the only accidents that should be addressed if the SAR is to serve as a reference for accidents for the site-specific application. Others that may be derived from a hazard analysis could include accidents resulting from operational error, instrument failure, lightning, and other occurrences. Accident situations that are not credible because of design features or other reasons should be identified and justified in the SAR.

(a) Cask Drop

The SAR should identify the operating environment experienced by the cask, as well as the drop events (i.e., end, side, corner) that could result. Generally the design basis is established either in terms of the maximum height to which the cask may be lifted when handled outside the reactor site spent fuel building or in terms of the maximum acceleration that the cask could experience in a drop.

(b) Cask Tipover

Although cask system supporting structures may be identified and constructed as being important to safety (i.e. designed to preclude cask tipovers), the NRC considers that cask tipover events should be analyzed. In some cases, cask tipover may be determined to be a credible hazard, and the associated analysis should reflect the conditions (e.g., heights and accelerations) associated with that hazard.

In the absence of an identified hazard, the NRC has accepted a non-mechanistic cask tipover about a lower corner onto a receiving surface from a position of balance with no initial velocity. The receiving surface for a horizontal or vertical drop may be either an unyielding hard surface; or, the receiving surface may be modeled as a reinforced concrete pad on an engineered fill[8,9]. The NRC has also accepted analysis involving the dropping of a cask with its longitudinal axis in the horizontal position that, with analysis of a vertical axis drop, could bound a non-mechanistic tipover case.

(c) Fire

The fire conditions postulated in the SAR should provide an "envelope" for subsequent comparison with site-specific conditions. The NRC accepts the methods discussed in 10 CFR Part 71.73. The NRC staff also accepts that the applicant may consider a fire based upon the limited availability of flammable material at an ISFSI (e.g., only that associated with vehicles transporting or lifting the cask or possibly nearby foliage). Regardless of which approach the applicant takes, the SAR should specify and justify the bounding conditions for a "design basis" fire

(d) Fuel Rod Rupture

The regulations require that the cask be designed to withstand the effects of accident conditions and natural phenomena events without impairing its capability to perform safety functions. Consequently, the NRC has asserted and the applicant should assume, during the cask analysis for conditions resulting from design-basis accidents and natural phenomena, a release of 100 percent of the initial rod fill gases and a release of 30 percent of the fission product gases from the fuel rods into the cask interior. The remaining 70 percent of the fission product gases are presumed to be retained within the fuel pellet.

(e) Leakage of the Confinement Boundary

Casks are designed to provide the confinement safety function under all credible conditions. Nevertheless, the NRC staff considers that, for assessment purposes and to demonstrate the overall safety of the storage cask system, the DCSS should be evaluated for the effects of a confinement boundary failure. The SAR should identify this failure as a bounding release caused by a non-mechanistic event and the effects should be evaluated as described in the Sandia National Laboratories Report 80-2124[10].

(f) Explosive Overpressure

The conditions under which an ISFSI may be exposed to the effects of an explosion vary greatly among individual sites. Generally, explosive overpressure is postulated to originate from an industrial accident. The effects of various sabotage methods on cask systems were evaluated separately by the Division of Fuel Cycle Safety and Safeguards in developing appropriate regulations in 10 CFR Part 73[11]. Therefore, explosive overpressures from sabotage events are not be considered in this SRP.

The extent to which explosive overpressure is addressed in the SAR directly affects the degree of site-specific review required. The principal concern in the SAR should be the effects of explosive overpressure on the storage system, rather than descriptions of hypothesized causes. Design parameters for blast or explosive overpressures should identify pressure levels as reflected ("side-on") overpressure, and should provide an assumed pulse length and shape. This discussion should provide sufficient information for licensees to determine if the effects of their site-specific hazards are bounded by the cask system design bases.

(g) Air Flow Blockage

For storage systems with internal air flow passages, the applicant should consider blockage of air inlets and outlets in an accident condition. The NRC staff considers that the effects of such an assumption should be utilized in determining the appropriate inspection intervals, and/or monitoring systems, for the DCSS.

(4) Natural Phenomena Events

The staff has generally considered that the following events should be evaluated in the SAR.

(a) Flood

The SAR should establish a design-basis flood condition. This condition may be determined on the basis of the presumption that the cask cannot tip over and the yield strength of the cask will not be exceeded. Alternatively, the SAR can show that credible flooding conditions have negligible impact on the cask design.

If the SAR establishes parameters for a design-basis flood, all of the potential effects of flood water and ravine flood byproducts should be recognized. Serious flood consequences can involve effects such as blockage of ventilation ports by water and silting of air passages. Other potential effects include scouring below foundations and severe temperature gradients resulting from rapid cooling from immersion.

(b) Tornado

The NRC staff accepts design-basis tornado wind loading as defined by RG 1.76 (Region 1)[12] and tornado missile impacts defined by NUREG-0800, Section 3.5.1.4[13]. Design criteria should be established for the cask on the basis of these wind loading and missile impact definitions. The cask should not tip over and that the capability to perform the confinement safety function should not be impaired. The NRC considers that tornados and tornado missiles may occur without warning. The review should note that in general, the effects of a tornado missile bound those of a light general aviation aircraft directly impacting a DCSS.

(c) Earthquake

The SAR should state the parameters of the DBE. For ISFSIs at reactor sites, this is equivalent to the SSE used for analysis of nuclear facilities, under 10 CFR Part 50. An analysis for an "Operating-Basis Earthquake" (OBE) is not required for an DCSS SAR prepared in accordance with 10 CFR Part 72. Cask tipover accidents are analyzed, but tipover caused by an earthquake may not be a credible event.

(d) Burial under Debris

Debris resulting from natural phenomena or accidents that may affect cask system performance may be addressed in the SAR or may be left to the site-specific application. Such debris can result from floods, wind storms, or land slides. The principal effect is typically on thermal performance.

(e) Lightning

Lightning typically has a negligible effect on cask systems; however, the requirements of the Lightning Protection Code and National Electric Code should be applied to the design of the cask system structures. These codes should be cited as part of the general design criteria for the cask system (see Section II.3.a, above). Lightning should also be addressed as a natural phenomenon in the SAR if cask system performance may be affected if lightning affects a component that is important to safety.

(f) Other

10 CFR Part 72 identifies several other natural phenomena events (including seiche, tsunami, and hurricane) that should be addressed for spent fuel storage. The SAR may include these as design-basis events or show that their effects are bounded by other events. If they are not addressed in the SAR and they prove to be applicable to a specific site, a safety analysis is required prior to approval for use of the DCSS under either a site specific, or general license.

3. Design Criteria for Safety Protection Systems

Because RG 3.61 does not distinguish between the principal design criteria that should be presented in Section 2 of the SAR and those that should be deferred to subsequent sections, the applicant may take one of several approaches. SAR Section 2 may discuss these criteria in general terms (similar to the wording in Section II.3 above), with details provided in later sections. Alternatively, SAR Section 2 may present detailed discussions of selected (or all) criteria. Past applicants have generally selected the latter approach. Subsequent chapters of this SRP provide detailed discussions of the design criteria applicable to each functional area (e.g. structural, thermal) without regard to those that may have been presented in SAR Section 2.

Cask system components that are to be used in facility areas subject to review under 10 CFR Part 50 should satisfy both the requirements in 10 CFR Part 72 (with review guided by this SRP) and 10 CFR Part 50 (with review guided by NUREG-0800 and applicable portions of RG 3.53[14]). Acceptance of the cask system in areas covered by 10 CFR Part 50 license requirements is not addressed in this SRP for approval under 10 CFR Part 72. If a reviewer knows that the cask system will be used at a specific reactor site, the NRR project manager for that site should be so informed. The reviewer is reminded that a likely matter of interest to NRR is heavy loads.

Regardless of where the descriptions and associated criteria are located in the SAR, reviewers should include a summary description and evaluation of the safety protection systems in the "Design Criteria" section of the SER. The system descriptions should address the functions of the various system components in providing confinement, cooling, subcriticality, radiation protection of the public and workers, and spent fuel retrieval. Summary criteria for the performance of the system as a whole in providing for these capabilities or functions should also be described and evaluated. Reviewers should verify that the design-basis assumptions presented are consistent with and reasonable for actual site or facility conditions.

Criteria relating to redundancy, and allowable levels of response by the DCSS under normal, off-normal, and design-basis conditions and events should be described and evaluated. In general, no unacceptable degradation in physical condition or functional performance should result from normal or off-normal conditions. The design criteria regarding limits of permissible system response and degradation resulting from a DBE should be evaluated against the SSC capabilities to perform the principal safety functions. Considerations of permissible responses should include detectability and corrective actions that may be proposed as conditions of system use.

Table 2-1 summarizes design criteria (and design bases) that should generally be identified during the initial stages of the review. This listing may vary depending on the details of the cask design.

(a) Continuous Monitoring

The Office of the General Counsel (OGC) has developed an opinion as to what constitutes "continuous monitoring" as required in 10 CFR Part 72.122(h)(4). The staff, in accordance with that opinion has concluded that both routine surveillance programs and active instrumentation meets the intent of "continuous monitoring". Cask vendors may propose, as part of the SAR, either active instrumentation or surveillance to show compliance with 10 CFR Part 72.122(h)(4).

The reviewer should note that some DCSS designs may contain a component or feature whose continued performance over the licensing period has not been demonstrated to staff with a sufficient level of confidence (e.g. rubber "O" rings). Therefore, staff may require the use of that active instrumentation, if the failure of that system or component causes an immediate threat to the public health and safety, and if that failure would not be detected by any other means. In some cases the vendor or staff in order to demonstrate compliance with 10 CFR Part 72.122(h)(4), may propose a technical specification requiring such instrumentation as part of the first use of a cask system. After first use and if warranted and approved by staff such instrumentation may be discontinued or modified.

The staff should verify that the applicant has met the intent of continuous monitoring so that they are able to determine when corrective action needs to be taken to maintain safe storage conditions.

VI. Evaluation Findings

Provide a summary statement similar to the following:

The staff concludes that the principal design criteria for the [cask designation] are acceptable with regard to meeting the regulatory requirements of 10 CFR Part 72. This finding is reached on the basis of a review that considered the regulation itself, appropriate regulatory guides, applicable codes and standards, and accepted engineering practices. A more detailed evaluation of design criteria and an assessment of compliance with those criteria as presented in Sections 3 through 14 of the SER.

Table 2-1 Outline of Design Criteria and Bases[a]

Design Criteria (Specify normal/off-normal/accident, if applicable)

Design Life (License restricted to 20 years)

Structural
    Design Code
        Containment (e.g., ASME[b], AISC[c])
        Non-containment
        Basket
        Trunnions
        Storage radiation and protective shielding and enclosure
        Transfer radiation and protective shielding and enclosure
        Cooling structure or system

    Design Weight

    Design Cavity Pressure   Normal/Off-Normal/Accident

    Response and Degradation Limits   Normal/Off-Normal/Accident

Thermal

    Maximum Design Temperatures
        Cladding   5-yr Cooled Fuel (As Applicable)
                                      10-yr Cooled Fuel

        Other Components

    Insolation
        Side/Top/Bottom

    Fill Gas

Confinement

    Method of Sealing

    Maximum Leak Rates
        Primary Seals
        Redundant Seals
        Cask Body

    Monitoring System Specifications

Retrievability

    Normal and Off-Normal

    After DBE and Conditions

Criticality

---

  [a]   This table should be filled out by reviewer for inclusion in the SER

  [b]   American Society of Mechanical Engineers

  [c]   American Institute of Steel Construction

Principal Design Criteria

Method of Control
    (Geometry, Fixed Poison, Borated Pool Water)

Minimum Boron Concentration
    Fixed
    Pool Water

Maximum k_{eff}

Burnup Credit (None currently permitted)

Radiation Protection/Shielding

Confinement Cask
    Surface   Normal/Off-Normal/Accident
    Position

Exterior of Shielding   Normal/Off-Normal/Accident
    Transfer Mode Position
    Storage Mode Position

ISFSI Controlled Area Boundary
    Normal/Off-Normal/Accident Dose Rate
    Annual Dose

Design Bases

Spent Fuel Specifications
    Type
    Configuration/Vendor
    Enrichment
    Weight or range of weights
    Burnup
    Type of Cladding
    Assemblies/Cask
    Dimensions

    Decay Heat/Assembly
        5-yr Cooled Fuel
        10-yr Cooled Fuel, etc.

    Gas Volume (@ Temperature)
    Fuel Condition/Damage Allowed
    Control Components

Normal Design Event Conditions

    Ambient Temperature
        Maximum
        Minimum

    Loading
        (Wet/Dry)

    Storage Handling Orientation
        (Vertical/Horizontal)

    Max lift height
    Other Conditions Considered in V.2.b.(1)

Off-Normal Design Event Conditions

    Summarize Events Considered in V.2.b.(2)

Design-Basis Accident Design Events and Conditions

End Drop	Lift Height (or Maximum Acceleration)
Side Drop	Lift Height (or Maximum Acceleration)
Tip-Over	Acceleration (if applicable)

Fire
    Duration
    Temperature

Other Events Considered in V.2.b.(3)
    (As Applicable)

Design-Basis Natural Phenomena Design Events and Conditions

    Flood
    Earthquake
    Tornado

    Other Events Considered in V.2.b.(4)
       (As Applicable)

VII. References

1. *U.S. Code of Federal Regulations*, "Licensing Requirements for the Independent Storage of Spent Nuclear Fuel and High-level Radioactive Waste," Part 72, Title 10, "Energy."

2. NUREG/CR-6407 INEL-95/0051 "Classification of Transportation Packaging and Dry Spent Fuel Storage System Components According to Importance to Safety," February 1996.

3. U.S. Nuclear Regulatory Commission, "Standard Format and Content for a Topical Safety Analysis Report for a Spent Fuel Dry Storage Cask," Regulatory Guide 3.61, February 1989.

4. NRC Bulletin 96-04: Chemical, Galvanic, or Other Reactions in Spent Fuel Storage and Transportation Casks, July 1996.

5. *U.S. Code of Federal Regulations*, "Reactor Site Criteria," Part 100, Title 10, "Energy".

6. NRC Bulletin 96-02: "Movement of Heavy Loads Over Spent Fuel, Over Fuel in the Reactor Core, or Over Safety Related Equipment", April, 1996

7. *U.S. Code of Federal Regulations*, "Packaging and Transportation of Radioactive Material," Part 71, Title 10, "Energy."

8. Drop Tests onto Concrete Pads for Benchmarking Response of Interim Spent Fuel Storage Installation, "Sandia National Laboratories, Albuquerque, New Mexico, September 1993.

9. "Low-Velocity Impact Testing of Solid Steel Billet onto Concrete Pads," Draft Summary of Twelve Drop Tests, Lawrence Livermore National Laboratory, Livermore, California, March 1, 1996.

10. E. L. Wilmot, Sandia National Laboratories Report "Transportation Accident Scenarios for Commercial Spent Fuel", SAND80-2124, TTC-0158, February 1981

11. *U.S. Code of Federal Regulations*, "Physical Protection of Plants and Materials," Part 73, Title 10, "Energy."

12. U.S. Nuclear Regulatory Commission, "Design Basis Tornado for Nuclear Power Plants," Regulatory Guide 1.76, April 1974.

13. U.S. Nuclear Regulatory Commission, "Standard Review Plan, Missiles Generated by Natural Phenomena," NUREG-0800, Section 3.5.1.4, July 1981.

14. U.S. Nuclear Regulatory Commission, "Applicability of Existing Regulatory Guides to the Design and Operation of an Independent Spent Fuel Storage Installation " Regulatory Guide 3.53, July 1982.

3.0 STRUCTURAL EVALUATION

I. Review Objective

In this portion of the dry cask storage system (DCSS) review, the NRC evaluates aspects of the DCSS design and analysis related to structural performance under normal and off-normal operations, accident conditions and natural phenomena events. In conducting this evaluation, the NRC staff seeks a high degree of assurance that the cask system will maintain confinement, subcriticality, radiation shielding, and retrievability of the fuel under all credible loads for normal and off normal accident conditions and natural phenomenon ("accident-level") events.

II. Areas of Review

This chapter of the DCSS Standard Review Plan (SRP) provides guidance for use in evaluating the design and analysis of the proposed cask system, with regard to its structural performance. All storage cask systems include a confinement cask that may have both internal components and integral external components. In addition, some cask systems have a variety of other components that are subject to NRC evaluation and approval.

Recognizing the diversity of the various cask system components, the NRC has broadly categorized the applicable review procedures and acceptance criteria, as follows:

- confinement cask
- reinforced concrete (RC) components
- other system components important to safety
- other components subject to NRC approval

Within these broad categories, the NRC focuses the DCSS structural evaluation, as described in Section V, "Review Procedures," using the following areas of review, as appropriate:

1. scope
2. structural design criteria and design features
   a. design criteria
      i. general structural requirements
      ii. applicable codes and standards
   b. structural design features
3. structural materials
4. structural analysis
   a. load conditions
      i. normal conditions
      ii. off-normal conditions
      iii. accidents
   b. structural analysis methods
      i. finite-element analysis
      ii. closed-form calculations
      iii. prototype or scale model testing
      iv. structural analysis of specific components
   c. structural evaluation
      i. summary structural capability
      ii. fabrication and construction
      iii. structural compatibility with functional performance requirements

III. Regulatory Requirements

1. Structures, systems, and components (SSC) important to safety must meet the regulatory requirements established in 10 CFR 72.24(c)(3) and (4), as well as 10 CFR 72.122(a), (b), and (c)[1].

2. Radiation shielding, confinement, and subcriticality must meet the regulatory requirements defined in 10 CFR 72.24(d); 10 CFR 72.124(a); and 10 CFR 72.236(c), (d), and (l).

3. As stated in 10 CFR 72.122(f) and (h)(l), the storage system design must allow ready retrieval of spent fuel without posing operational safety problems.

4. As stated in 10 CFR 72.102(f), the design-basis earthquake (DBE) must be equal to or greater than the safe-shutdown earthquake (SSE) of nuclear plant sites previously evaluated under 10 CFR Part 100[2] or, in the case of sites licensed before the implementation of 10 CFR Part 100, developed under Topic III-2 of the Systematic Evaluation Program (SEP)[3].

5. As stated in 10 CFR 72.24(c) and 10 CFR 72.236(g), the analysis and evaluation of the structural design and performance must demonstrate that the cask system will allow storage of spent fuel for a minimum of 20 years with an adequate margin of safety.

6. Reinforced concrete structures may have a role in shielding, form ventilation passages and weather enclosures, and providing protection against natural phenomena and accidents. The pertinent regulations include 10 CFR 72.24(c) and 10 CFR 72.182(b) and (c).

IV.   Acceptance Criteria

The most important function of the structural analysis is to ensure sufficient structural capability for every applicable section of the cask system to withstand the worst-case loads under accident conditions and natural phenomena events. Withstanding such loads enables the cask system to successfully preclude the following negative consequences:

- unacceptable risk of criticality
- unacceptable release of radioactive materials
- unacceptable radiation levels
- impairment of ready retrievability

Because of the diversity of cask system components that are subject to NRC evaluation and approval, it is inconceivable that staff would be able to define objective structural review criteria that address all possible component configurations. Moreover, no single structural code, (such as the ASME B&PV[4]) covers the design of all spent fuel storage systems. Consequently, the acceptability of any given structure will be contingent upon a combination of adherence to applicable portions of multiple codes and a review of the functional performance of the structure taken as a whole. This combined approach allows the designer to request relief and the reviewer to impose additional restrictions when warranted by specific design features.

In general, the DCSS structural evaluation generally seeks to ensure that the proposed design and analysis fulfill the following acceptance criteria, which reflect the industry codes and standards that the NRC staff has accepted in past DCSS structural evaluations:

With exceptions for the confinement cask, ANSI/AND-57.9[5] generally applies to the design and construction of an independent spent fuel storage installation (ISFSI). Table 3-1 includes extracts of ANSI/ANS-57.9 that apply to the design and construction of ISFSI structures other than the confinement system.

1. Confinement Cask

a. Steel Confinement Cask

The structural design, fabrication, and testing of the confinement system and its redundant sealing system should comply with an acceptable code or standard, such as Section III of the Boiler and Pressure Vessel Code (B&PV)[6] promulgated by the American Society of Mechanical Engineers (ASME). (The NRC has accepted use of either Subsection NB or Subsection NC of this code.) Other design codes or standards may be acceptable depending on their application.

    i.   The NRC staff evaluates the proposed limitations on allowable stresses and strains in the confinement cask, reinforced concrete components, system components important to safety, and other components subject to review, by comparison with those specified in applicable codes and standards. Where certain proposed load combinations will exceed the accepted limits for localized points on the structure, the applicant should provide adequate justification to show that the deviation will not affect the functional integrity of the structure.

ii. The NRC has accepted the use of applicable subsections of the ASME B&PV Code, Division 1, for components used within the confinement cask but not integrated with it. This includes the "basket" structure used in casks to restrain and position multiple fuel elements.

b. Concrete Containments

i. ACI 359[7] (also designated as Section III, Division 2, of the ASME B&PV Code, Subsection CC) constitutes an acceptable standard for prestressed and reinforced concrete that is an integral component of a radioactive material containment vessel that must withstand internal pressure in operation or testing.

ii. If ACI 359 pertains to a given ISFSI structure, it applies to all aspects of the design, material selection, fabrication, and construction of that structure. The NRC has not accepted the proposed substitution of elements from ACI 318[8] or ACI 349[9] for any portion of ACI 359 with regard to the structure of an ISFSI. ISFSI structures to which ACI 359 applies shall also meet the minimum functional requirements of ANSI/AND-57.9 for subject areas not specifically addressed in ACI 359.

2. Reinforced Concrete (RC) Structures Important to Safety, but not within the Scope of ACI 359

The NRC accepts the use of ACI 349 for the design, material selection and specification, and construction of all reinforced concrete structures that are not addressed within the scope of ACI 359. However, in such instances, the design, material selection and specification, and construction must also meet any additional or more stringent requirements given in ANSI/AND-57.9, as incorporated by reference in NRC Regulatory Guide (RG) 3.60[10]. Section V of this chapter provides additional guidance regarding specific review procedures.

3. Other Reinforced Concrete Structures Subject to Approval

The NRC accepts the use of either ACI 318 or ACI 349 for reinforced concrete structures that are subject to approval but are not important to safety. Section V of this chapter provides additional guidance regarding specific review procedures.

4. Other System Components Important to Safety

The NRC accepts the use of ANSI/AND-57.9 (together with the codes and standards cited therein) as the basic reference for ISFSI structures important to safety that are not designed in accordance with the Section III of the ASME B&PV Code. However, both the lifting equipment design and the devices for lifting system components that are important to safety must comply with American National Standards Institute (ANSI) Standard N14.6[11].

The NRC accepts the load combinations shown in Table 3-1 for structures not designed under either Section III of the ASME B&PV Code or ACI 359. These load combinations are based upon ANSI/AND-57.9, with supplemental definition of terms and combinations.

The principal codes and standards include the following references that may apply to steel structures and components:

a. American Institute of Steel Construction (AISC), "Specification for Structural Steel Buildings — Allowable Stress Design and Plastic Design"[12]

b. AISC, "Load and Resistance Factor Design Specification for Structural Steel Buildings"[13]

c. American Welding Society, "Structural Welding Code Steel," AWS D1.1[14]

   d. American Society of Civil Engineers, "Minimum Design Loads for Buildings and Other Structures," ASCE 7[15] [however, note that load combinations established on the basis of ANSI/AND-57.9 (DCSS SRP Table 3-1) are to be used.]

   e. ACI 349-85, Appendix B[16], for embedments or 10.14 for composite compression sections, as applicable, when constructed of structural steel embedded in reinforced concrete.

5. Other Components Subject to NRC Approval

For structural design and construction of other components subject to NRC approval, the principal codes and standards include the following:

   a. ASCE 7

   b. Uniform Building Code[17] (UBC)

   c. AISC, "Specification for Structural Steel Buildings—Allowable Stress Design and Plastic Design"

   d. AISC "Code of Standard Practice for Steel Buildings and Bridges"[18]

   e. ASME B&PV Code, Section VIII[19]

V. Review Procedures

In evaluating the structural design and performance of a proposed DCSS, select and emphasize aspects of the following review procedures, as appropriate for the particular DCSS, in relation to the acceptance criteria summarized in Section IV, above:

● Description of Structures, Systems, and Components Important to Safety

Verify that the applicant's safety analysis report (SAR) clearly identifies the proposed structural design and construction of SSC that are important to safety and necessary for effective functional performance and safety of the DCSS. Review the SAR and supplemental material submitted by the applicant to assess compliance with the applicable scope and content requirements defined in 10 CFR 72.24 or 72.230. (Focus in particular on requirements and conditions of use related to design, construction, implementation, operation, and maintenance of structural SSC. (10 CFR 72.28 requires applicants to propose conditions of use in license applications for approval under Subpart L.) Request any additional information required from the applicant at an early stage in the review process.

● Applicable Codes, Standards, and Specifications

NRC guidelines recommend that the safety evaluation report (SER) prepared by the NRC staff include a table (in the design criteria evaluation section) summarizing the applicable reference sources. This table should identify all source documents cited in the SAR, their usage (e.g., description of model, prior NRC approval of cask system elements, design code, construction code), and acceptability for that usage is recommended. The sources of interest include documents directly referenced in the SAR; sources of material incorporated by reference; and codes, standards, specifications, and other sources of criteria that further define the design and construction of the proposed structures. If not tabulated, the consolidated review and assessment of reference sources should otherwise be included in the SER.

● Loads and Load Combinations

Verify that the loads and load combinations are as specified in Chapter 2, of this SRP. If the applicant has not adequately justified any deviations from the acceptance criteria for loads and load combinations, identify the deviations as unacceptable and transmit them to the applicant for further justification.

The SAR should include a comprehensive table of load combinations and safety margins for selected structural sections of components important to safety (or otherwise subject to NRC evaluation); this table should be included in the SER. The summary table should include sufficient structural sections and forms of loadings (e.g., shear, flexure, axial, and combined stress situations) to verify that the lowest margins of safety are represented for the various components. In addition, this table can be used to summarize the structural capacity evaluation.

- Design and Analysis Procedures

  Determine that the applicant's design and analysis procedures and assumptions are conservatively defined on the basis of accepted engineering practice. Review the behavior of the structure under various loads, and the manner in which these loads are treated in conjunction with other coexistent loads; assess compliance with the acceptance criteria defined in Section III of this chapter.

- Structural Acceptance Criteria

  Review the proposed limitations on allowable stresses and strains in the confinement cask, reinforced concrete components, system components important to safety, and other components subject to review. Compare the proposed limitations with those specified in the applicable codes and standards. Where the applicant proposes to exceed the accepted limits for certain load combinations at localized points on the structure, evaluate the justification provided to ensure that the deviation will not affect the functional integrity of the structure. If the justification is not acceptable, request additional justification and bases.

- Materials, Quality Control, and Special Fabrication Techniques

  Review the information provided in the SAR regarding materials, quality control programs, and special fabrication techniques, if any, and compare the proposal with the acceptance criteria in Section II of this chapter. If the applicant proposes to use a new material not addressed in prior approvals, the applicant must provide sufficient test and user data to establish the acceptability of the material. Similarly, review and evaluate any new quality control programs or construction techniques to ensure that they will not degrade the structural quality, integrity, or function of the DCSS.

- Testing, and In-Service Surveillance Requirements

  Review the proposed pressure test procedures for the confinement cask by comparison with the procedures described in ASME Code, Section III, Subsection NB-6000. Also review the proposed acceptance test and maintenance requirements for trunnions by comparison with those described in the ASME Code and ANSI N14.6, as applicable. Review any other proposed testing and in-service surveillance programs on a case-by-case basis. Also review SAR Section 9 to verify that the applicant has include all appropriate acceptance tests, and address all required evaluations in Section 9 of the SER.

- Conditions for Use of Structures

  Review the structural evaluation to determine if conditions of use or technical specifications (or "license conditions") should be associated with the structural design or proposed fabrication and construction. Review the appropriateness of and need for any proposed technical specifications related to structural design and construction. Determine whether any additional technical conditions related to structural performance are needed and, if so, provide input to the conditions of use discussed in the SER. Also describe the basis for the suggested conditions in the structural evaluation section of the SER.

  Structure-related conditions of use may be linked to evaluations performed under other sections (such as a field verification that maximum concrete temperatures predicted from thermal analysis will not be exceeded). 10 CFR 72.44(c)(4) suggests a minimum structural license condition. Similarly, 10 CFR 72.234(a) suggests a minimum structural condition of approval for certification; this standard also incorporates 10 CFR 72.236(j) by reference to 72.236.

The remainder of this section provides specific review procedures for each of the four categories of cask system components, including the confinement cask, reinforced concrete components, other safety-related system components, and other components subject to NRC approval. Within each of these broad categories, the specific review procedures focus the DCSS structural evaluation using the areas of review identified in Section II of this chapter.

1. Confinement Cask

The structural review of the confinement cask addresses drawings, plans, sections supporting computations, and specifications for those structural components comprising confinement barriers. The review also addresses structural and sealing interfaces and connections that are necessary to complete the confinement system (as defined in 10 CFR Part 72). In addition, this review includes evaluation of components that serve no structural function, in order to confirm that they do not impair the functioning of the confinement cask.

a. Scope

The SAR must describe all SSC important to safety in sufficient detail to allow evaluation of their structural effectiveness. In addition, the SAR must identify all codes and standards applicable to SSC important to safety.

The discussion in the SAR must demonstrate that all SSC important to safety will be designed and fabricated to quality standards commensurate with the importance to safety of the function to be performed. In addition, SSC important to safety must be designed to accommodate the combined loads anticipated during normal, off-normal, accident, and natural phenomenon events with an adequate margin of safety.

b. Structural Design Criteria and Design Features

i. Design Criteria

The NRC generally considers the following design criteria to be acceptable to meet the structural requirements of 10 CFR Part 72:

(1) General Structural Requirements

The proposed cask must maintain confinement of radioactive material under normal and off-normal operations, accident conditions, and natural phenomenon events. In addition, neither the cask nor any basket within the cask may deform under credible loading conditions in a manner that would jeopardize the subcritical condition or retrievability of the fuel.

The design must adequately protect the spent fuel cladding against gross rupture caused by degradation resulting from design or accident conditions. In addition, the design must ensure that the spent fuel will not experience accelerations that would damage its structural integrity or jeopardize its subcritical condition or retrievability.

The applicant must analyze the cask to show that it will not tip over or drop in its storage condition as a result of a credible natural phenomenon event. A tipover or drop is always assessed as a bounding condition during handling operations.

Radiation shielding in the cask system is required to protect the public and workers at the ISFSI, and such shielding must not degrade under normal or off-normal conditions or events. The shielding function may degrade as a result of a design-basis accident (e.g., loss of liquid neutron shielding resulting from a drop accident). However, the loss of function must be readily visible, apparent, or detectable. (Any permissible degradation in shielding must be shown to result in dose rates sufficiently low to permit recovery of the damaged cask, including unloading if necessary). In addition, the procedures specified in the SAR for use after such a DBA should include procedures for testing the effectiveness of the shielding.

(2) Applicable Codes and Standards

The structural design, fabrication, and testing of the confinement system and its redundant sealing system should comply with acceptable code or standards. Use of codes and standards previously accepted by the NRC expedites the evaluation process. Use of other codes and standards, definition of criteria composed of extracts from multiple codes and standards with overlapping scopes, or substitution of other criteria, in whole or in part, in place of acceptable published codes or standards requires a custom NRC review and may delay the evaluation process.

An accepted code for design, fabrication, and test of steel confinement casks is Section III of the ASME B&PV Code. (Specifically, the NRC has accepted use of either Subsection NB or NC.) Other design codes or standards may be acceptable depending on their application. The NRC has accepted use of the applicable subsections of the ASME Code for cask system components used within the confinement cask but not integrated with it. This includes the "basket," which is a structure used in casks to restrain and position multiple fuel elements.

The NRC has also accepted applicable subsections of Division 1, of the ASME Code, for structural external integral elements of the confinement (e.g., Subsection NF for integral supports).

The NRC accepts use of Regulatory Guides 7.11[20] and 7.12[21] as bases for determining the potential for brittle fracture. These regulatory guides also incorporate a portion of NUREG/CR-1815[22] by reference.

The applicant may define the fatigue limits of the cask structural materials on the basis of the provisions of Reference 3 or the guidance provided in Regulatory Guide 7.6[23]. However, since casks are typically subjected to non-cyclic loads, fatigue may not be a significant concern.

ii. Structural Design Features

Review the cask-related descriptive information presented in SAR Section 1, as well as any related information provided in SAR Section 3. The drawings, figures, tables, and specifications included in the SAR should fully define the structural features of the cask. These may include the cask body (including an inner shell, an outer shell, and a lead gamma shield), inner and outer lids and bolts, port covers and bolts, vent port covers to be welded in place, neutron shields and shell, trunnions, fuel basket, and impact limiters (if used).

Coordinate with the confinement review (Chapter 7 of this SRP) to verify that the SAR clearly identifies the confinement boundaries. These boundaries include the primary confinement vessel; its penetrations, seals, welds, and closure devices; and the redundant sealing system. Ensure that the applicant has provided proper specifications for all welds and, if applicable, that the applicant has adequately designed and specified the bolt torques for closure and locking devices.

Review the list of weights and calculation of the cask center of gravity. Verify that the applicant used the appropriate limiting cases in the structural evaluations.

Review the cask structural materials that are in direct contact with each other to verify that they will not produce a significant chemical or galvanic reaction and the attendant corrosion or combustible gas generation.

Review confinement boundary weld designs for compliance with the design code used for the confinement boundary. Acceptable requirements appear in ASME Code Section III, Subsections NB-3352 and NC-3352, "Permissible Types of Welded Joints," and NB-4240 and NC-4240, "Requirements for Weld Joints in Components."

The NRC has previously accepted alternative confinement boundary weld designs (such as NB-5200 or NC-5200, typically for Category C welded joints). These acceptable alternatives achieve equivalent structural integrity, but do not meet all the provisions of NB-3352 or NC-3352 for full penetration welds, or do not meet the NDE requirements for full volumetric nondestructive examination. The NRC has also accepted alternative designs for the welds of the head or flat end plate to the cylindrical portion of the confinement vessel. However, the NRC has required the alternative designs to include redundant welds to provide redundant sealing of the confinement systems.

In addition, welds must be well-characterized on drawings using standard welding symbols and/or notations, as discussed in American Welding Society (AWS) Standard A2.4[24].

c. Structural Materials

The information provided on structural materials must be consistent with the application of accepted design criteria, codes, standards, and specifications selected for the storage cask system. For example, if the applicant elects to use design criteria from Section III of the ASME B&PV Code, the materials selected for the cask must be consistent with those allowed by the ASME Code subsection related to design. Acceptable requirements include the ASME-adopted specifications given in Section II, Part A, "Ferrous Metals"; Part B, "Nonferrous Metals"; Part C, "Welding Rods, Electrodes, and Filler Metals"; and Part D, "Properties."

In reviewing the structural materials, consider the sources of information; properties used in the structural evaluation (including those that affect performance under both static and dynamic loadings for normal, off-normal, and accident conditions and natural phenomenon events); and suitability for the proposed life of the ISFSI. Preferred sources include industry and Government codes, standards (including NUREG-3760[25]), and specifications. Review the applicability and acceptability of all other sources, such as manufacturer's test data and handbooks. Published articles, research reports, and texts have generally not been accepted by the NRC as primary sources of information concerning material properties.

The intent of this portion of the DCSS structural evaluation is to determine the acceptability of all materials that have a structural role in confinement system structures and other structures important to safety (e.g., the basket, impact limiters, and shielding). However, this review should also include evaluating the suitability of the materials for the proposed structural and operational application, as well as the material properties that may affect structural design and evaluation over the approved period of use. For example. The reviewer should be familiar with the information contained in NRC bulletin 96-04 "Chemical, Galvanic, or other Reactions in Spent Fuel Storage and Transportation Casks"[26]

The reviewer must consider the suitability of materials to be used in Structures Systems and Components (SSC) important to safety. The material properties and characteristics needed to satisfy these functional safety requirements must be maintained over the 20-year approval period. For some components, the life cycle may include conditions experienced during cask fabrication, loading, transport, emplacement, storage, transfer, retrieval, and decommissioning. Service conditions include normal, off-normal operations, accidents, and natural phenomena events.

Where historical data are available, they may furnish reasonable assurance that the material is suitable for a given component, provided that the service conditions are sufficiently similar to those of the precedent. Where such analogies are not available, the knowledge, judgement and experience of the reviewer must be used to ensure the use of good engineering practice. When additional information is required, a brief literature search may suffice. When necessary, the required information may be requested of the applicant.

Analyze the potential for corrosion and ensure that the applicant established and used appropriate corrosion allowances for the structural analyses. Also consider the static and dynamic (where appropriate) stresses and the limits used for the structural design and evaluations.

When dissimilar metals are connected electrically, a galvanic cell is established in which electrochemical interactions are enhanced. For example, if bolts are anodic to a large component of a cask system, the bolts may corrode quickly and impair their ability to function successfully as a fastener in the cask system. The galvanic series lists metals in terms of their electrochemical potential, which is a parameter that may be useful in establishing the likelihood of problems in either aqueous systems or vapors of moderate to high humidity. Because different metals may be used within a cask system, it is important to note the possible interactions between dissimilar-metal systems and to evaluate the possibilities for unfavorable interactions in relation to functions that are important to the safety of the systems. For example, in the presence of a large ferrous cathodic surface area, zinc will corrode at a rapid rate. The products of this reaction are gaseous hydrogen, ions and zinc compounds. These reaction products must be tolerable and they must not impair any safety function.

Additional material requirements apply for structural designs governed by the ASME B&PV Code, Section III, Subsection NB or NC. Specifically, these requirements include examination before fabrication, testing and analyses, and traceability. In particular, the SAR must acknowledge compliance with the requirements of the following Section III paragraphs, or their equivalent:

- NB-2121 or NC-2121, "Permitted Material Specifications"

- NB-2130 or NC 2130, "Certification of Material"

- NB-2500 or NC-2500 "Examination and Repair of Pressure-Retaining Material"

- NB-2400 or NC 2400, "Welding Material"

A DCSS serves to confine spent fuel and maintain safe storage conditions throughout its service life. Construction Codes, e.g. ASME B&PV Code Section III, give reasonable assurance that the as-fabricated material will provide the necessary integrity. It is noted that the ASME Code Section III applies specifically to maintaining pressure boundaries and supporting structures in nuclear power plants. It may not necessarily be applicable to all DCSS. However, designers may choose to cite it as the code to which selected components are to be fabricated. Codes such as the ASME B&PV are not likely to address all the potential performance problems (e.g. cracking, creep, corrosion, etc.) which may arise from environmental, electrochemical or dynamic-loading. These and other effects are specific to the individual application and are frequently, outside the intended application of the code. Thus, even where codes have been judiciously applied, the reviewer must establish that sufficient background, experience and knowledge exists to provide reasonable assurance on the long-term performance.

For a material that is not normally welded many questions must to be answered to ensure that the process chosen for fabrication will yield a durable component. Cracking problems with weldments are numerous and expert advise or appropriate research and development may be warranted for a new welding application.

The reviewer should ensure that bolts are properly heat treated. Improper heat treatment may result in bolt cracking either under normal conditions (if tempered too little) or under off-normal (accident) conditions (if tempered too much).

The SAR should also include tables detailing material properties and allowable stresses and strains (as appropriate).

A list of all materials used and the proposed service conditions for those materials, during loading, storage, and unloading is a useful aid during the review. A table of this type is included here as Appendix B. This table illustrates various types of information that the reviewer needs from an application, to aid in determining the suitability of the materials for the service conditions. It includes the name and safety classification of each component part of the dry cask storage system and, where applicable, the function , the material specification(s) to which it is produced, and the nominal values for the following parameters: strength, surface finish or coating, materials (if dissimilar) with which it is in direct contact. If welded, the list includes the welding process and filler metal. Other tabulations include the stress (nominal and maximum) in service, the residuals (chemicals/foreign matter) on the surface of the component after loading and after storage, the service temperatures (for the storage period, during loading and during unloading), the internal pressure (min, max) and the type/composition of gas or liquid in the container. The tabulation should include all materials used for components with an important-to-safety function, e.g. confinement, transport, criticality control, shielding. In addition, materials that coat or in other ways support or interact (physically, chemically, or electrochemically) with the important-to-safety materials should be tabulated. Information in this table can aid the reviewer to formulate the types of performance-related questions that are important for each component of a storage system.

Verify that the properties used are appropriate for the load conditions of interest (e.g., static or dynamic, impact loading, hot or cold temperature, wet or dry conditions). Review SAR Section 12 to ensure that the applicant considered any appropriate restrictions regarding temperature or environmental conditions for the materials. Verify that the SAR clearly references acceptable sources of all material properties.

Coordinate with the thermal review to determine the appropriate temperatures at which allowable stress limits should be defined. For most cask materials, the stress limits should be defined at the maximum temperature for each material, as established by the SAR thermal analysis.

Materials that function as neutron absorbers and gamma shields should be fabricated from materials that can perform well under conditions of service that are appropriate for these components over the 20 year licensing period. Coordinate with the criticality and shielding reviews to ensure that during storage and accident conditions the materials do not creep or slump to an extent that impairs the capability to perform its safety function

Ensure that the applicant considered the potential for brittle fracture, especially for cask system components that may be subject to impact during exterior handling and transfer operations. The potential for brittle fracture of some components important to safety has resulted in conditions of use that preclude transfer operations under extremely low temperature conditions. Ensure that any assumptions about internal heat generation for the brittle fracture analysis are defined on the basis of the maximum storage life and the possibility of a partial load in the cask. Verify that SAR Section 12 addressed any necessary restrictions regarding cask handling at low temperatures, and that these restrictions are addressed in Section 12 of the SER.

If the cask has impact limiters, the applicant should thoroughly test and verify their nonlinear impact characteristics. In addition, the applicant should tabulate and describe the crush characteristics and properties of the limiters in the directions that are to be used.

d. Structural Analysis

i. Load Conditions

To meet the structural requirements of 10 CFR Part 72, the DCSS design must accommodate the full spectrum of load conditions, including all anticipated normal, off-normal, and accident-level conditions (including natural phenomenon events). The system should not experience any deformation or loss of safety function capability under normal operating conditions. However, the system may experience some deformation, but no loss of safety function capability, in response to accident.

(1) Normal Conditions

Normal conditions and events are those associated with cask system operations, including storage of nuclear material, under the normal range of environments. The SAR should state the assumed limits of normal use environments, in order to support evaluation of cask system suitability for use at a specific site.

Loads normally applicable to a confinement cask include weight, internal and external pressures, and thermal loads associated with operating temperature. The loads experienced may vary during loading, preparation for storage, transfer, storage, and retrieval operations. The weight is the maximum or design weight (including tolerances) of the cask as it is stored and loaded with spent fuel. However, depending on the operation and procedures, the weight should also include water fill. The applicant should evaluate all orientations of the cask body and closure lids during normal operations and storage conditions, including loads associated with loading, transfer, positioning, and retrieval of the confinement cask.

Internal pressures result from hydrostatic pressure, cask drying and purging operations, filling with non-reactive cover gas, out-gassing of fuel, refilling with water, radiolysis, and temperature increases. Temperature variations and thermal gradients in the structural material may cause additional stresses in the cask and closure lids. Coordinate with the thermal review (Chapter 4 of this SRP) to determine the conservative (or enveloping) values and combinations of the cask internal pressures and temperatures for both hot and cold conditions. Use the temperature gradients calculated in SAR Section 4 to determine thermal stresses. Note that if the confinement system has several enclosed areas, all areas may not have the same internal pressures. In some casks, enclosed areas consist of the cask cavity and the region between the inner and outer lids.

Required evaluations include weight plus internal pressures and thermal stresses from both hot and cold conditions. Verify that the applicant included the maximum thermal gradient, as determined in the thermal analysis, when evaluating thermal stresses.

(2) Off-Normal Conditions

The review should identify and evaluate all off-normal events and conditions described in Chapter 11 of this SRP. Review the off-normal conditions and events for those that affect the confinement cask structure. The confinement cask components should satisfy the same structural criteria required for normal conditions, as discussed above.

The SAR should clearly identify anticipated off-normal conditions and events that may reasonably be expected to occur during the life of the cask system at the proposed site. In addition, the SAR should state the environmental limits to support comparison of the cask system design bases with specific site environmental data. Off-normal conditions and events can involve potential mishandling, simple negligence of operators, equipment malfunction, loss of power, and severe weather (short of extreme natural phenomena).

(3) Accident-Level Events and Conditions

Follow the guidance below in reviewing the structural response to accident conditions. Note that the SAR *must* address *at a minimum* each of the following accidents. However, this discussion may not address all of the potential events or accidents that apply to a cask (Chapter 11 of this SRP addresses the identification and evaluation of accidents.)

(a) Cask Drop and Tipover

The SAR should identify the operating environment experienced by the cask and the drop events (end/side/corner) that could result. Generally, applicants establish the design basis in terms of the maximum height to which the cask is lifted outside the spent fuel building, or the maximum deceleration that the cask could experience in a drop. The design-basis drops should be determined on the basis of the actual potential handling and transfer accidents.

Drops of a cask with axis vertical onto an edge may involve subsequent rotation. Drops with the axis generally vertical should be analyzed for the both conditions of a flush impact and an initial impact at a corner of the cask, in recognition that the worst-case loadings for the contents of the cask (versus damage to the cask itself) may result from different orientations at impact.

Applicants should analyze cask tipover regardless of the credibility of occurrence. The NRC will accept cask tipover about a lower corner onto a hard receiving surface from a position of balance with no initial velocity. The NRC has also accepted analysis of cask drops with the longitudinal axis horizontal, which together with analysis of a drop with axis near vertical, could bound a non-mechanistic tipover case.

Until recently, NRC staff has accepted an unyielding surface for determining the bounding cask deceleration loads which can far exceed the decelerations experienced by a cask dropping onto or tipping over the concrete storage pad that will bend and deform. As described in a latter section, prototype or scale model testing can be used to obtain more realistic cask deceleration or equivalent load for quasi-static analyses. Alternatively, applicants can develop an analytical model to calculate cask deceleration loads. In the analytical approach, the hard receiving surface for a drop or tipover accident need not be an unyielding surface and its flexibility may be included in the modeling. However, the analytical model should be validated. The staff has completed a series of low-velocity impact tests of steel billets, and is in the process of developing detailed guidance for using the billet test results to validate a cask-pad-soil interaction model for predicting cask deceleration loads.

(b) Explosive Overpressure

Explosion-induced overpressure and reflected pressure may result from explosion hazards associated with explosives and chemicals transported by rail or on public highways, natural gas pipelines, and vehicular fires of equipment used in the transfer of casks. Explosions may result from detonation of an air-gaseous fuel mixture. With the exception of transfer vehicle accidents, the explosion hazards are typically similar to those for facilities subject to reviews under 10 CFR Part 50[27]. Note that this explosive overpressure differs from that associated with the design-basis radiological sabotage event. The combination of physical security planning and cask design is intended to protect the public against such a threat.

The review for site-specific explosion hazards would be left for the license application for the specific site if explosions are not addressed in the SAR. Alternatively, the SAR can state the level of overpressure, reflected pressure, and/or pressure differentials assumed to result from an explosion; this level would then serve as the quantitative envelope for future comparison with hazards for specific site installations. The pressure criteria for the assumed design-basis wind or tornado may also serve as an envelope for the explosive pressures, for comparison with actual site hazards.

If the SAR includes bounding explosion effects for which the cask system is to be approved, verify that the applicant also provided structural analyses of those effects for cask system structures that may be affected. The SAR should identify the maximum response determined. That response should be sufficiently low such that while damage may occur it would not impair the capability of the component to perform its safety functions. In addition, the SAR should identify any post-event inspection and remedial actions that may be necessary.

(c) Fire

Chapter 4 of this SRP addresses potential fire conditions. Fire-related structural evaluation considerations include increased pressures in the confinement cask, changes in material properties (e.g., temporary loss of strength at elevated temperatures and permanent loss of strength because of annealing), stresses caused by different coefficients of thermal expansion and/or temperatures in interacting materials, and physical destruction (e.g., surfaces of concrete exposed to intense or prolonged high temperatures).

Review and evaluate the discussion in the SAR concerning the treatment of structural effects associated with the presumed fire. Evaluate the appropriateness of the applicant's analysis of those structural effects for the assumed parameters of the design-basis fire. Confirm that the applicant defined the confinement cask pressure capacity on the basis of the cask material properties at the temperature resulting from the fire.

The NRC has accepted the fire parameters included in 10 CFR Part 71[28] as the basis for characterizing the heat transfer associated with fire during storage. Spalling of concrete that may result from a fire is generally considered acceptable and need not be estimated or evaluated. Such damage is readily detectable, and appropriate recovery or corrective measures may be presumed. The NRC accepts concrete temperatures that exceed the temperature limits of ACI 349 for accidents, provided that the temperatures result from a fire. However, corrective actions may need to be taken for continued safe storage.

(d) Flood

Review the applicant's evaluation of the cask system design with regard to the structural consequences of a flood event. The SAR may stipulate an assumption that the cask system not be used at any site where there is potential for flooding. In this case, the cask would have to be placed on a reactor site at a location above the maximum probable flood. (SAR Section 12 should state this condition.) Alternatively, a license application for a site with flooding potential would require a full analysis.

One possible structural consequence of a flood is that a vertically stored cask may tip over or translate horizontally (slide) because of the water velocity. Another possible consequence is that external hydrostatic pressure will exceed the capacity of the cask. The applicant may state the critical water velocity and hydrostatic pressure as bounds for the SAR flood analysis.

The NRC accepts application of the requirements of ANSI/ANS-57.9, Section 6.17.4.1, to the flood case for overturning and sliding of stored confinement casks and other cask system structures (with a safety factor of 1.1 for accidents cases). The applicant should state the basis for estimation of lateral pressure on a structure as a result of water velocity.

The NRC accepts the use of Hoerner's *Fluid-Dynamics Drag*[29] for estimating drag coefficients and net lateral water pressure. An approach for calculating the velocity corresponding to the cask stability limit is to assume that the cask is pinned at the outer edge of the cask bottom, that the cask rotates about that outer edge, and that the pinned edge does not permit sliding. The overturning moment from the velocity of the flood water can be compared to the stability moment of the cask (with buoyancy considered). The structural consequences of the flood event are typically bounded by analyses for the drop or tipover accident cases.

Review the analysis of the confinement cask for flood-related hydrostatic pressure. The analysis should include the combined effects of weight, external hydrostatic pressure, internal pressure(s), and thermal stress. Resistance of the confinement cask to flood-related hydrostatic pressure should be analyzed in accordance with Section III, Subsection NB or NC of the ASME B&PV Code (depending on the subsection used for design).

Additional flood consequences include potential scouring under a foundation, damage to access routes, temporary blockage of ventilation passages with water, blockage of ventilation passages and interstitial spaces between the confinement cask and shielding structure with mud, and steep temperature gradients in the shielding structure and confinement cask. While the consequences of these conditions may be analyzed in the SAR, the licensee should consider these factors when siting an ISFSI.

    (e) Tornado Winds

Verify that the SAR addresses the potential structural consequences of design-basis tornado or extreme wind effects. Review the load combination analyses for acceptable inclusion of tornadoes and tornado missiles.

Confinement casks may be vulnerable to overturning and/or translation caused by the direct force of the drag pressure while in storage or during transfer operations. ANSI/ANS-57.9 provides acceptable criteria for resistance to overturning or sliding.

Confinement casks are generally not vulnerable to damage from overpressure or negative pressure associated with tornadoes or extreme winds. However, they may be vulnerable to secondary effects, such as wind-borne missiles (see (f), below) or collapse of a weather enclosure. Tornadoes or extreme winds have been a governing load condition in previous reviews for major structures that form part of an ISFSI system. (These structures may provide for shielding, cooling paths, and/or transfer and storage operations.)

Tornadoes typically produce the greatest "design-level" wind effects for American sites. However, there are some potential American sites at which high winds may be more severe than the credible tornado. The SARs for a limited set of potential sites could reflect high wind effects as a basis for structural analysis. If the certificate is to include proven resistance to tornadoes or extreme winds, the SAR documentation must identify the wind levels (e.g., in miles or kilometers per hour), source (tornado or high wind), and specific wind-driven missiles (shape, weight, and velocity) for which the design is to be evaluated.

Regulatory Guide 1.76[30] provides applicable tornado-related parameters. The NRC accepts the use of ASCE 7 for conversion of wind speed to pressure and for typical building shape factors. Conversion of tornado or other wind speeds to pressure in the SAR documentation should assume that the cask system is at sea level. In addition, the SAR should cite the source for drag coefficients used to compute net forces on objects. (Hoerner's *Fluid-Dynamics Drag* is one acceptable source.)

For the design tornado wind pressure, the NRC accepts use of the pressure derived from conversion of wind speed, without gust or importance factors, for tornadoes. If the design-basis wind is caused by extreme winds, the NRC accepts the computational approach given in ASCE 7 for determining pressures. This approach adds gusts, importance, exposure, and height above ground to the analysis. The computational approach of ASCE 7, has also been accepted for normal and off-normal wind loadings.

Tornadoes and high winds can produce a significant negative pressure differential between interior spaces and the outside. This is a function of wind speed and factors relating to the structure. The magnitude of negative pressure depends on other parameters of the tornado or wind, and on wall pressure coefficients (as expressed in ASCE 7). There is no need for the SAR to separately state negative pressure to establish an envelope for approval since negative pressure is insignificant with regard to confinement cask accident pressure analysis.

The NRC does not accept the presumption that there will be sufficient warning of tornadoes that operations such as transfer between the fuel pool facility and storage site may never be exposed to tornado effects. Overturning during onsite transfer is considered by the staff to be a design-basis event. The tornado analysis should determine if tornado-induced overturning is bounded by drop and tipover cases. In addition, the SAR should show that the cask system will continue to perform its intended safety functions (criticality, radioactive material release, heat removal, radiation exposure, and ready retrievability).

(f) Tornado Missiles

Review the applicant's evaluation of the cask system design with regard to the structural consequences of wind-driven missile impact. (Regulatory Guide 1.76 and NUREG-0800[31] describe the effects of tornado missiles.) The SAR should define the missile parameters for which the cask system is to be evaluated. Among the possible missile effects, the SAR should address those that may result in a tipover, and those that may cause physical damage as a result of impact. The damage should not result in unacceptable radiation dose or significantly impair either criticality control, heat removal, or the ready retrievability of the fuel.

The NRC has accepted use of the analytical approaches given in ORNL-NSIC-5, Volume 1, Chapter 6[32], for estimating the potential effects of missile impact on steel sheets, plates, and other structures. Further guidance on analytical acceptable approaches for use in ISFSI design is provided in NUREG-0800, Section 3.5.3, "Barrier Design Procedures." In addition, for analysis and design regarding the ability of reinforced concrete structures to resist missiles, the NRC has accepted use of R.P. Kennedy's "Review of Procedures for the Analysis and Design of Concrete Structures to Resist Missile Impact Effects"[33].

Cask systems are not required to survive missile impacts without permanent deformation. However, the maximum extent of damage from a design-basis event must be predicted and should be sufficiently limited. Moreover, the capability of the SSC to perform their safety functions should not be impaired.

(g) Earthquake

Review the applicant's evaluation of the cask design with regard to the structural consequences of the earthquake event. As explicitly stated in 10 CFR Part 72, the design-basis earthquake (DBE) must be no less than the safe shutdown earthquake for a reactor at sites that have been evaluated under Appendix A to 10 CFR Part 100. Cask designs must satisfy the load combinations that encompass earthquake, including those for sliding and overturning in ANSI/ANS-57.9, Section 6.17.4.1). The applicant should demonstrate that no tipover or drop will result from an earthquake. In addition, impacts between casks should either be precluded, or should be considered an accident event for which the cask must be shown to be structurally adequate.

The SAR documentation should include analysis of the potential for impacts between components of the cask system. These could include contact between the confinement shell and its inner components or outer shield, and the rocking and fall back of a vertically or horizontally oriented confinement cask on its supports.

Cask systems are not required to survive a DBE without permanent deformation. However, the maximum extent of damage from a design-basis event must be predicted, and the capability to provide principal safety functions should not degrade.

ii. Structural Analysis Methods

Review the applicant's structural analysis, stresses, and stress combinations resulting from different loads. Look for satisfactory evidence that the applicant properly used acceptable analytical approaches and tools. In addition, the applicant should have performed and reviewed the associated computations internally under an acceptable independent design review (equivalent to ANSI N 45.11) and quality assurance procedures. The scope of the staffs review does not necessarily include performing detailed parallel computations (such as finite element analyses) to validate submitted computations or their results. The reviewer may perform separate, less extensive calculations when these could most readily evaluate suspected problems.

The applicant's analysis of stresses and stress combinations resulting from different structural loads should be consistent with the subsection of Section III of the ASME B&PV Code used in designing the component.

For Class 1 and Class 2 components, respectively, Subsection NB or NC of the ASME B&PV Code, defines the requirements for categorizing stresses and determining allowable stress limits for confinement casks. These references also provide definitions of stress categories and stress intensity limits for normal and off-normal operating conditions. For level D or accident conditions, Appendix F to the ASME B&PV Code provides definitions of the stress intensity limits.

In accordance with these references, stress intensity is defined on the basis of the maximum shear stress theory for ductile materials. Since the maximum shear stress is not identical to the maximum octahedral shear stress, octahedral shear stresses should not be compared with the stress intensity limits. Values for the stress intensity limits are defined in Appendices I and III of the ASME Code. Stresses resulting from inertial and pressure loads should be considered primary stresses since they can be shown to be self-limiting. Thermal stresses resulting from temperature gradients may be considered secondary stresses if they are self-limiting and do not cause structural failure.

(1) Finite-Element Analyses

Because of the complexity of many structural design considerations, load conditions, and structural design computations are often performed using finite-element analysis.

The applicant should perform the finite-element analyses using a general-purpose program that is well benchmarked and widely used for many types of structural analyses. Codes, such as SCANS and CASKS may be used as confirmatory tools, but are not applicable for primary analyses in the SAR because they have simplifying assumptions regarding cask geometry, materials, and structural behavior.

When possible, solutions from finite-element analyses should be compared with closed-form calculations. While they are unlikely to exactly duplicate the complex load conditions analyzed with the finite element program, they can verify simpler portions. For example, the formulas for the stress in a cylinder with end-caps can be used to check the stress state caused by internal pressure in the cask.

To be consistent with the provisions in Section III of the ASME Code, the analyses should use linear material properties. For materials that do not serve in a structural capacity (such as shielding materials), inelastic material properties may be used for cask components that are not stress-limited and respond inelastically to the load conditions for storage casks. The SAR should identify the sources used for the inelastic material properties.

Lead shielding, which is typically not stress-limited, can be modeled either with elastic or inelastic properties. The elastic modulus and limit used for lead in the elastic analysis should be determined on the basis of the potential temperature of the material. An appropriate plasticity model of lead can be used to account for its inelastic behavior.

Nonstructural components of the confinement cask are generally not included in finite element models. However, the models should include any influence these nonstructural components may have on the structural performance of the cask. Possible influences include the nonstructural components' inertial weight, restraint to motion of the structural components, and localized influence on load applications because of geometrical effects.

Bolted connections can be modeled either discretely or with contact conditions. To discretely model the bolted connections, the applicant should use appropriate element types and material properties. With contact conditions, the interfaces joined by the bolts can be modeled as tied.

The number of discrete finite elements used in the model should reflect the type of analysis being performed. That is, regions in the model of high stress or displacement should have a higher number of elements than regions that have a nearly equilibrated state of stress or are in a uniform stress field. Consequently, the applicant should conduct sensitivity studies to determine the appropriate number of nodes or elements for a particular model.

(2) Closed-Form Calculations

The applicant should perform closed-form calculations for relatively simple structural load conditions or conditions for which a formula has been developed. Closed-form calculations are also typically used to check the results of finite-element analyses. In addition, this type of calculation can be used for analyses involving principles of conservation of energy and comparisons of overturning moments.

One source of closed-form equations accepted by the NRC is *Formulas for Stress and Strain* (Roark 1965)[34]. Use of a particular equation or formulation for the load conditions should be justified as appropriate. The most important aspect of the calculations to evaluate is the basis for the assumptions used in the calculations. In many cases, the calculations are faulty in that they fail to include portions of the cask or the load conditions are idealized.

To be consistent with the provisions in Section III of the ASME Code, the analyses should use linear material properties. Linear analysis should be the basis for all closed-form calculations.

(3) Prototype or Scale Model Testing

Applicants may perform prototype or scale model testing in lieu of, or to supplement impact analysis for cask drop conditions. However, use of scale model testing to directly demonstrate that the cask design meets the regulatory requirements may be difficult, because leakage rates and other radiological limits may not correspond to the same scaling factor used for the model. Consequently, impact tests intended to independently demonstrate regulatory compliance are usually prototypical and may be used to assess performance of a specific component like an impact limiter. Applicants should perform a sufficient number of tests to cover all design impact conditions and other uncertainties (such as the hardness of the receiving target surface).

Applicants can also perform drop tests to obtain an equivalent static load to be used for a quasi-static analysis of the cask. Drop tests can also yield key data, such as the spring stiffness of the target surface, which may then be used to perform a dynamic analysis of the cask[35].

A scale model must properly simulate the distribution of the loads (weights), the geometry (dimensions), and the material properties of the cask. If the scale model omits any parts of the cask, the applicant should provide adequate justification and should discuss the resulting effects on test findings. In addition, the applicant should develop a test plan to identify the test conditions, the parameters to be measured during and after the test, and the test acceptance criteria.

(4) Structural Analysis for Specific Cask Components

The following paragraphs present a few specific examples of structural analysis for some of the confinement cask components:

(a) Trunnions

Review the design of the trunnions, their connections to the cask body, and the cask body in the local area around the trunnions. The design of the trunnions can be either non-redundant or redundant. In either case, the design should meet the requirements of ANSI N14.6 for critical loads and the requirements of NUREG-0612[36].

Non-redundant lifting systems should be designed for not less than 6 times the material yield strength and 10 times the material ultimate strength given the design lift weight of the loaded cask. Redundant lifting systems should be designed for not less than 3 times the material yield strength and 5 times the material ultimate strength given the design loaded lift weight of the cask. (Acceptance testing requirements for trunnions are discussed in Chapter 9 of this SRP.)

For a typical trunnion design, the maximum stress occurs at the base of the trunnion as a combination of bending and shear stresses. A conservative technique for computing the bending stress is to assume that the lifting force is applied at the cantilevered end of the trunnion and that the stress is fully developed at the base of the trunnion. If other assumptions are used, the applicant should provide adequate justification. In addition, the applicant should evaluate the stresses and forces in the trunnion connections with the cask body and in the cask body near the trunnions.

(b) Fuel Basket

Review the fuel basket design to assess the applicant's analysis of the combined effects of weight, thermal stresses, and cask-drop impact forces. The weight supported by the basket should be the maximum or design weight of the spent fuel to be stored. In addition, the applicant should evaluate all credible potential orientations of the cask and basket during cask drop. End or side drops typically produce the greatest structural demand on various basket components. In the end drop, the basket is supported by the bottom of the confinement cask cavity upon impact. In the side drop, the basket structure and points of contact with the confinement cask must support the mass of the basket and loaded fuel.

In previous DCSS evaluations, the NRC has accepted two approaches for analyses regarding the structural capability of basket to acceptably survive cask drop. The first approach uses dynamic analyses in a two-step process. In step 1, the applicant performs a dynamic analysis of the cask body impacting a target surface, and assesses the response of the cask body to determine the maximum response from the cask drop impact. This maximum response can then be translated into a forcing function, which can be applied to the supporting contact points of an appropriate model of the fuel basket.

The second approach uses a quasi-static analysis of the basket subjected to the equivalent acceleration inertial load derived from the cask-drop impact analysis. In this analysis, the applicant should apply the equivalent acceleration inertial load, using an appropriate model of the basket with the location(s) most vulnerable to the impact. Support provided by the inside surface of the cask cavity should be represented by the appropriate boundary conditions on the outside edge of the basket. In addition, the applicant should conservatively select the equivalent acceleration inertial load such that it bounds the possible inertial loads resulting from a cask-drop accident onto the bounding target surfaces. If applicable, the inertial load should also account for dynamic amplification effects by using a dynamic amplification factor.

The applicant should also evaluate the buckling capacity of the cask basket materials. Acceptable guidance for this evaluation is provided in Section III of the ASME B&PV Code and NUREG/CR-6322[37]. For this evaluation, the applicant should select the appropriate end conditions used in the buckling capacity equations on the basis of sensitivity studies. These studies can bound the range of conditions, which are typically either fixed for a welded connection or free if there is no rigid connection.

(c) Closure Lid Bolts

Review the design analysis for the closure-lid bolts to ensure that it properly includes the combined effects of weight, internal pressure(s), thermal stress, O-ring compression force, cask impact forces, and bolt pre-load. Typically, applicants specify the pre-load and bolt torque for the closure bolts on the basis of bolt diameter and the coefficient of friction between the bolt and the lid. Externally applied loads (such as the internal pressure and impact force) produce direct tensile force on the bolts, as well as an additional prying force caused by lid rotation at the bolted joint. The tensile bolt force obtained by adding together the pressure loads, impact forces, thermal load, and O-ring compression force should then be compared with the tensile bolt force computed from the pre-load and operating temperature load alone. The larger of the two calculated tensile forces should control the design. The maximum design bolt force should then be obtained by combining the larger direct tensile bolt force with the additional prying force. The weight is derived from the maximum or design weight of the closure lids and any cask components supported by the lids. Acceptable analytical methods for closure bolts are given in NUREG/CR-6007[38].

Review the bolt engagement lengths. If the lids are fabricated from relatively non-hardened materials, threaded inserts may be used in the closure lids to accommodate the hardened material of the bolts.

iii. Structural Evaluation

(1) Structural Capability

Review the applicant's structural analyses to assess the tables or statements regarding margins of safety or compliance with ASME Code stress limits, overturning, and other criteria. The comparisons of capability versus demand for the various applicable loading conditions should be presented in the same terms used in the design code (e.g., type of stress). In addition, margins of safety should be included on the basis of comparisons between capacity and demand for each of structural component analyzed. The minimum margin of safety for any structural section of a component should be included for the different load conditions.

(2) Fabrication and Construction

The NRC has accepted fabrication of confinement casks in accordance with Section III of the ASME B&PV Code. If the fabrication, construction, or assembly deviate in any way from the subsection of this standard used for design, the SAR must explicitly state the applicant's justification for the deviation, and the justification must be acceptable to the NRC.

In reviewing the fabrication and construction of the confinement cask, focus especially on any specifications regarding preparation for welding, materials to be used in the welds, performance of welding, and inspection of welds that do not fully comply with Section III of the ASME B&PV Code.

Welding procedure qualifications and welding performance qualifications should conform with the requirements of Section IX of the ASME B&PV Code[39]. For confinement welds, the SAR documentation should include the bases for detailed welding procedure specifications (WPSs) that identify acceptable ranges of essential welding variables (listed in Section IX of the ASME Code for all approved welding processes). The welding variables should be recorded as quality assurance records during production runs. All welds should be performed by pre-qualified personnel in accordance with written procedures.

Testing of weld integrity may involve a combination of ASME-approved weld test techniques, which do not necessarily result in full radiographic examination, but some volumetric inspection (e.g., ultrasonic testing (UT)) may be necessary.

(3) Structural Compatibility with Functional Performance Requirements

Review the SAR documentation to confirm that the design of the cask structure provides for satisfactory functional performance. This includes operating suitability within specified limiting conditions and satisfaction of the basic safety criteria under all credible events and environmental conditions.

The SER should clearly identify the confinement system and other structures important to safety, each of which should have sufficient structural capability for every applicable section to withstand the worst-case loads under accidents and conditions, to successfully preclude the following:

- unacceptable risk of criticality

- unacceptable release of radioactive materials to the environment

- unacceptable radiation dose to the public or workers

- significant impairment of ready retrievability of stored nuclear materials

This position does not necessarily require that all confinement system and other structures important to safety survive all design-basis accidents and extreme natural phenomena without any permanent deformation or other damage. Some load combination expressions for the DBE and conditions for structures important to safety permit stress levels that exceed yield. The SAR should include computations of the maximum extent of potentially significant transient deformations and any permanent deformations, degradation, or other damage that may occur. Verify that the applicant has performed computations, analyses, and/or tests and that both the tests and results they are acceptable to the NRC in order to clearly demonstrate that any permanent deformations, degradation, or other damage that may occur does not render the system performance unacceptable.

Structures important to safety are not required to survive accidents to the extent that they remain suited for use for the life of the cask system without inspection, repair, or replacement. If the life of structures important to safety may be degraded by accident conditions, there must be SAR commitments and procedures for determining and correcting the degradation, and performing other acceptable remedial action.

Review the proposed technical specifications to ensure that they include adequate restrictions on cask handling and operations to preclude the possibility of damage to the structure or the confined nuclear material. Operating controls and limits of the technical specifications (reviewed under Chapter 12 of this SRP) should be included in both the SAR and the SER, and should describe actions to be taken and inspections to be conducted upon occurrence of events that may cause such damage.

2. Reinforced Concrete Components

This section presents guidance and review procedures for conducting structural evaluations regarding the reinforced concrete components of the cask system. Specifically, the reinforced concrete structures subject to NRC evaluation include SSC that are to be included in the approved cask system. These may be of concern because of their safety function or importance to safety (per 10 CFR 72.24 (c)).

a. Scope

Reinforced concrete structures may play multiple roles in providing radiological shielding or forming ventilation passages, weather enclosures, structural supports, access denial, foundations, earth retention, anchorages, floors, walls, movable shields, bulk fill, and protection against natural phenomena and accidents. Bulk fill may be emplaced within an enclosing structure to provide shielding or strength.

Reinforced concrete structures may be cast at the site, or cast elsewhere. reinforced concrete structures may also comprise *combinations* of cast-in-place and precast sections that are assembled by bolting, welding, fitting, grouting, or placing additional concrete at the site. They may also include concrete cast as part of a composite confinement cask with metallic liner. However, this subsection does not address the metallic liner of a composite confinement cask, its closures, or its internal components.

Embedments and attachments to reinforced concrete structures are analyzed as parts of the reinforced concrete structure unless they are specifically addressed elsewhere in Chapter 3 of this SRP. Embedments and attachments are considered to include components that are cast or grouted into the reinforced concrete structure, inserts, embedded pipes and conduits, or lightning protection and grounding systems.

b. Structural Design Criteria and Design Features

i. Design Criteria

(1) General Structural Requirements

All concrete used in storage cask system ISFSIs, and subject to NRC review, should be reinforced, regardless of the functional role or need for structural strength or integrity. The concrete specifications should state the reinforced concrete design code or standard applicable to its intended use and acceptable to the NRC.

The structural design of the reinforced concrete structures shall withstand the effects of credible accident conditions and natural phenomenon events without impairing their capability to perform safety functions. The principal safety functions include maintaining subcriticality, containing radioactive material, providing radiation shielding for the public and workers, and maintaining retrievability of the stored fuel.

The NRC has not required that exterior reinforced concrete pavements used for vehicular traffic, parking, or equipment access to the ISFSI storage area be designed as important to safety. Moreover, the SRP does not address the design or evaluation of pavements that are not considered structurally integral with the foundation of an reinforced concrete cask system structure that is subject to review. Nonetheless, reinforced concrete aprons that extend from a structure and are structurally integral with the structure are also elements of the foundation. As such they should be reviewed for compliance with the same code applicable for the attached reinforced concrete structure. If a pavement incorporates points for fastening supports that are important to safety (as may be used for transfer operations) that section of the pavement necessary for the function should be designed as a foundation in accordance with ACI 349.

Reinforced concrete pads that support confinement casks in storage do not constitute "pavements." As such, they should be designed and constructed as foundations under an applicable code, such as ACI 318, ACI 349 or UBC. Such pads typically are not classified as important to safety; however, in some cases they may be.

The applicant should consider the potential for liquefaction or other soil instabilities attributable to vibrating ground motion, and the pad should be designed with this in mind. Inspection Procedure 60851[40] and Regulatory Guide 3.60[41] provides guidance regarding soil engineering and seismic analysis requirements.

Steel embedments in reinforced concrete structures must satisfy the requirements of the design code applicable to the reinforced concrete structure. Similarly, structural steel must satisfy the requirements of the applicable steel design code.

(2) Applicable Codes and Standards

Review the codes and standards identified in the SAR, as well as their proposed applications. This subsection addresses the codes and standards that the NRC has accepted for reinforced concrete ISFSI structures, categorized by application (i.e., concrete containments, reinforced concrete structures important to safety but not within the scope of ACI 359, other reinforced concrete structures subject to NRC approval, and steel attachments to reinforced concrete structures).

ANSI/ANS-57.9 generally applies to ISFSI design and construction (with exceptions for confinement casks). Table 3-1 includes extracts from ANSI/ANS-57.9 that are particularly applicable to reinforced concrete structure design and construction. The table also includes corresponding evaluation guidance for use in reviewing the SAR documentation.

The NRC has not accepted the use of a set of criteria selected from multiple standards and codes, except when the selected criteria meet the most limiting requirements of each code. However, in recognizing a graded approach to quality assurance, the NRC has approved the use of ACI 349 for design and material selection for reinforced concrete structures important to safety (not confinement), but has allowed the optional use of ACI 318 as an alternative standard for construction, as described in this subsection.

Note that codes other than those discussed herein (e.g., the Electric, Life Safety, and Lightning Protection Codes[42] promulgated by the National Fire Protection Association (NFPA)) may apply to the design and construction of the cask system. It is acceptable to include such codes in the design by inclusion in the SAR. Where designs of structures subject to approval are also covered by such other codes, the review should include evaluation of compliance with those codes.

(a) Concrete Containments

ACI 359, also designated Section III, Division 2, of the ASME Boiler and Pressure Vessel Code, Subsection CC, is acceptable for prestressed and reinforced concrete that is an integral component of a radioactive material containment vessel that must withstand internal pressure in operation or testing. ACI 359 should be applied on the basis of containment function, regardless of whether the concrete structure is fixed or portable and regardless of where the concrete structure is fabricated. ACI 359 also applies to structural concrete supports constructed as an integral part of the containment.

If ACI 359 applies to an ISFSI structure, it applies to the entire design, material selection, fabrication, and construction of that structure. The NRC has not accepted the substitution of elements of ACI 349 or ACI 318 for any portion of ACI 359 for an ISFSI structure. In addition, ISFSI structures for which ACI 359 applies shall also meet the minimum functional requirements of ANSI/ANS-57.9, where ACI 359 does not include requirements regarding the specific subject area.

(b) Reinforced Concrete Structures Important to Safety, But Not Within the Scope of ACI 359

The NRC accepts the use of ACI 349 for the design, material selection and specification, and construction of all reinforced concrete structures that are not addressed within the scope of ACI 359. However, in such instances, the design, material selection and specification, and construction must also meet any additional or more stringent requirements given in ANSI/ANS-57.9, as incorporated by reference in RG 3.60.

The following paragraphs identify the portions of ACI 349 and ASTM standards that apply to design (including material selection) and must be met by applicants who choose to use ACI 318 for construction. (The paragraph references are as in ACI 349-90.) Unlisted and excepted sections address construction requirements, for which the NRC accepts substitution of ACI 318.

Chapter 1,	"General Requirements," Sections 1.1 and 1.5 (except references to construction), and Sections 1.2 and 1.4
Chapter 2,	"Definitions"
Chapter 3,	"Materials" (except Sections 3.1, 3.2.3, 3.3.4, 3.5.3.2, 3.6.7, and 3.7)

Chapter 4,	"Concrete Quality," Section 4.1.4
Chapter 6,	"Form Work, Embedded Pipes, and Construction Joints," Sections 6.3.6(k) and 6.3.8
Chapter 7,	"Details of Reinforcement"
Chapter 8,	"Analysis and Design General Considerations"
Chapter 9,	"Strength and Serviceability Requirements" (but see 2.2.d(1), below)
Chapter 10,	"Flexure and Axial Loads"
Chapter 11,	"Shear and Torsion"
Chapter 12,	"Development and Splices Information"
Chapter 13,	"Two-way Slab Systems"
Chapter 14,	"Walls"
Chapter 15,	"Footings"
Chapter 16,	"Precast Concrete"
Chapter 17,	"Composite Concrete Flexural Members"
Chapter 18,	"Prestressed Concrete"
Chapter 19,	"Shells"
Appendix A,	"Thermal Considerations"
Appendix B,	"Steel Embedments" (but note that the load combinations and variation requirements of ANSI/ANS-57.9 must be met in addition to those of ACI 349, Section 9.2, cited at Section B.3.2
Appendix C,	"Special Provisions for Impulsive and Impactive Effects" (except that the load combinations and variation requirements of ANSI/ANS-57.9 must be met in addition to those of ACI 349, Section 9.2

In addition, the following ASTM standard specifications apply to design and material specification (as referenced in ACI 349-90) and are acceptable to the NRC for design and construction of reinforced concrete structures:

A-36, A-53, A-82, A-184, A-185, A-242, A-416, A-421, A-496, A-497, A-500, A-501, A-572, A-588, A-615, A-706, A-722, C-33, C-144, C-150, C-595, and C-637[a]

(c) Other reinforced concrete Structures Subject to NRC Approval

The NRC accepts use of either ACI 318 or ACI 349 for reinforced concrete structures that are subject to NRC approval but are not important to safety. If ACI 349 is used for design, the NRC accepts use of ACI 318 for construction. The NRC also accepts the following as criteria as an alternative to the temperature requirements of ACI 349, A.4, but only for the specified uses and temperature ranges:

1. *If concrete temperatures of general or local areas are 200 °F in normal or off-normal conditions/ occurrences, no tests to prove capability for elevated temperatures or reduction of concrete strength are required.*

2. *If concrete temperatures of general or local areas exceed 200 °F but would not exceed 300 °F, no tests to prove capability for elevated temperatures or reduction of concrete strength are required if Type II cement is used and aggregates are selected which are acceptable for concrete in this temperature range. The following criteria for fine and coarse aggregates are acceptable:*

    a. *Satisfy ASTM C33 requirements and other requirements referenced in ACI 349 for aggregates, and*

    b. *Have demonstrated a coefficient of thermal expansion (tangent in temperature range of 70 °F to 100 °F) no greater than 6x10-6 in./in./°F, or be one of the following minerals: limestone, dolomite, marble, basalt, granite, gabbro, or rhyolite.*

3. *If concrete temperatures of general or local areas in normal or off-normal conditions or occurrences do not exceed 225 °F, the requirements of 1 and 2, above, apply to the coarse*

---

[a] Note that this list does not include A-616, A-617, A-767, A-775, or C-989, which are listed in ACI 318. These standard specifications apply if ACI 318 is used for construction.

aggregate, but fine aggregate that meets 1, above, and is composed of quartz sands or sandstone sands may be used in place of compliance with 2.

(d) Steel Attachments to reinforced concrete Structures

Codes and standards applicable for steel attachments to reinforced concrete structures are described in Subsection IV.3 for structures important to safety and in Subsection IV.4 for other structures subject to approval.

ii. Structural Design Features

Review the adequacy of the information provided in the SAR documentation regarding the physical design of reinforced concrete structures. This should include the following as a minimum:

- dimensioning of all surfaces

- locations, sizes, configuration, spacing, welding, enclosure (e.g., spirals, stirrups), and depth of cover of reinforcement

- locations and specifications for control, contraction, and construction joints

- materials, with defining standards or specifications

- review information on the physical design of embedments and attachments. This should include the following as a minimum:

  - locations, configuration, depth of embedment, interfaces; material; connections and connectors; and, protective or functional coatings

  - dimensions, materials, and specifications for welds

c. Structural Materials

i. Reinforced Concrete Components

Review the completeness, accuracy, and acceptability of the identification and stated properties of the reinforced concrete component materials.

Materials and material properties used for design and construction of reinforced concrete structures within the scope of ACI 359 must comply with the descriptions and requirements of that standard.

Materials and material properties used for the design and construction of reinforced concrete structures important to safety but not within the scope of ACI 359 should comply with the requirements of ACI 349.

Materials and material properties used for the design and construction of reinforced concrete structures that are not important to safety, but are subject to approval should comply with the requirements of ACI 318 (or ACI 349 if that code is used for design of the structures).

ii. Embedments and Attachments

Review the completeness and acceptability of the identification and stated properties of the material to be used for embedments, inserts, conduits, pipes, or other items that are to be embedded in the concrete. Embedments must satisfy the requirements of the code used in designing the reinforced concrete structure in which they are embedded (e.g., ACI 359, ACI 349, or ACI 318). Aluminum should not be used for any embedded objects that will be in contact with wet concrete (because of the potential for concrete degradation from an adverse chemical reaction).

Review the completeness and acceptability of the identification and stated properties of the material to be attached to the reinforced concrete structures. The material must satisfy requirements appropriate to its importance to safety. Unless otherwise specified in this SRP, steel structural attachments must comply with the appropriate requirements of ACI-349.

d. Structural Analysis

i. Load Conditions

Subsection V.1.d, above, provides guidance regarding the review of load conditions applicable to ISFSI structures in general. This subsection focuses on load conditions of special concern, and load combinations specifically for reinforced concrete structures. Review the appropriateness, completeness, and correctness of the applicant's proposed implementation of these load conditions and combinations for the reinforced concrete structures.

Load definitions and load combinations shown in Table 3-1 have been accepted by NRC for analysis of steel and reinforced concrete ISFSI structures important to safety. The load combinations are as included or derived from ANSI/ANS 57.9 and ACI 349.

Structures important to safety should have sufficient capability for every section to withstand the worst-case normal and off-normal conditions without permanent deformation and with no degradation of capability to withstand any future loadings.

(1) Normal Conditions

Review the SAR documentation to ensure adequate inclusion of the following conditions that may be of particular concern for reinforced concrete structures:

- live and dynamic loads associated with transfer of the confinement cask to and from its storage position

- live and dynamic loads associated with installing closures

- load or support conditions associated with potential differential settlement of foundations over the life of the cask system

- thermal gradients associated with the normal range of operations and ranges of ambient temperature

- thermal gradients that may result from impingement of rain on highly heated concrete

(2) Off-Normal Conditions

Review the SAR to ensure adequate inclusion of the following off-normal operations and events that may be of particular concern for reinforced concrete structures:

- live and dynamic loads associated with equipment or instrument malfunctions, or accidental misuse during transfer of the confinement cask to and from its storage position

- situations in which a confinement cask is jammed or moved at an excessive speed into contact with a reinforced concrete structure.

- the impact of reinforced concrete structures by a suspended transfer, confinement, or storage cask

- off-normal ambient temperature conditions (Although they may be less severe than accident conditions, these may be of concern because of different sets of factors in the off-normal and accident load combinations, and because concrete temperature limits for off-normal conditions are the same as for normal conditions. Note that greatly elevated concrete temperatures are allowed for accident conditions, in accordance with ACI 349, Section A.4.)

(3) Accident Conditions and Natural Phenomena events

Review the SAR for adequate inclusion of the following conditions associated with accident and conditions that may be of special concern for reinforced concrete structures:

- loads associated with accidental drops or other impacts during transfer of the confinement cask to and from its storage position

- events that produce extreme thermal gradients in the concrete

- contact caused by earthquake between the confinement cask and the reinforced concrete structures

- drop of a closure into position or onto the structure

The ACI codes are intended to ensure ductile response beyond initial yield of structural components. ACI 349 also imposes conditions on design (beyond those of ACI 318) that effectively increase ductility. In particular, review the proposed reinforced concrete design to ensure that it provides code levels of ductility, by satisfying of the pertinent ACI 349 provisions. Seismic loads are considered to be "impulsive" and, therefore, are subject to the additional design constraints of Appendix C to ACI 349. Other accident conditions or natural phenomenon events may also produce impulsive or impactive loadings requiring the additional requirements of Appendix C to ACI 349.

Check the steel reinforcement schedules and drawings to ensure that any reinforcing steel quantities, sizes, and locations are consistent with the design analysis. Use of more shear and enclosing reinforcement (e.g., stirrups, ties, and spirals) than required does not reduce ductility for the member. Constraints regarding the use of excess steel to ensure that ductility is not reduced do not apply to the shear and enclosure reinforcement.

In particular, consider the following aspects of the design:

- upper limit (60,000 psi, 4219 kgf/cm^2) on the specified yield strength of reinforcement, and lower limit (3000 psi, 211 kgf/cm^2) on concrete specified compressive strength (f'c)

- limit on the amount (cross-section area) of compressive reinforcement in flexural members

- requirements on continuation and development lengths of tensile reinforcement

- specifications for confinement and lateral reinforcement in compression members, in other compressive steel, and at connections of framing members

- aspects of the design that ensure flexure controls (and limits) the response

- requirements for shear reinforcement

- limitations on the amount of tensile steel in the flexural members relative to that which would produce a balanced strain condition

- projected maximum responses to design-basis loads within the permissible ductility ratios for the controlling structural action

- embedments designed to fail in the steel before pullout from the concrete

In addition, review the construction specifications or descriptions (to the extent included in the SAR documentation) to ensure that substitution of materials, use of larger sizes, or placement of larger quantities of steel will be precluded; and that provisions for splicing or development of reinforcing steel will not reduce ductility of the members.

ii. Structural Analysis Methods

Review the analytical documentation regarding the structural analysis methods used for design and verification of the reinforced concrete structures. In particular, ensure that the structural analysis of structures within the scope of ACI 359 comply with the requirements of that standard.

The NRC accepts strength design as presented in the current revision of ACI 349 for reinforced concrete structures important to safety that are not within the scope of ACI 359. If the applicant uses another design approach, the review conducted within the scope of the DCSS SAR evaluation should include in-depth comparison of that approach with the provisions of ACI 349.

The NRC accepts the use of procedures and approaches that are applicable to an ISFSI as described by the regulations referenced in Regulatory Guide 3.53[43]. The NRC also accepts the use of guidance in NUREG-0800 for analysis of natural phenomena; however, the load combinations shown in Table 3-1 and the design and construction requirements of the codes cited above take precedence. For estimation of wind, snow, and rain loads, and for conversion of tornado wind speed to pressure, the NRC accepts ASCE 7. Similarly, the NRC accepts ASCE 4[44] and ASCE 7 as the standards for seismic analysis. In addition, the NRC accepts tornado missile impact analysis in accordance with Kennedy's *Review of Procedures for the Analysis and Design of Concrete Structures to Resist Missile Impact Effects.*

(1) Strength Design

Strength (or "ultimate strength") design is the approach usually used in American reinforced concrete design. Strength design is the only design approach that has been accepted for ISFSI reinforced concrete structures not within the scope of ACI 359, and it is the approach used in the current revisions of ACI 318 and ACI 349. These design codes were developed on the basis of extensive empirical experience with concrete construction. The current strength design approach as presented in these codes includes empirically derived requirements and constraints. Determination that a reinforced concrete structure designed by another approach satisfies ACI 349 typically requires clause-by-clause review of the code for compliance.

(2) Allowable Stress Design

Allowable stress design was formerly used as the basis for ACI codes related to reinforced concrete design. However, those codes do not reflect additional experience gained through observations of structural performance and experimental testing, which has since been included in the current approach to strength design. A clause-by-clause comparison of the structural design for compliance with ACI 349 should be performed for reinforced concrete structures not designed using ACI 349.

(3) Analytical Codes and Models

The NRC has accepted the use of different analytical codes and models for structural analysis of reinforced concrete structures. Uses have included development of stresses resulting from seismic events and thermal gradients. The NRC does not require use of computer models and codes for analysis of the responses or stresses of simple ISFSI concrete structures. In addition, the NRC does not require that the codes used have been developed under rigid nuclear safety quality controls (e.g., ASME NQA-2[45]). However, the codes must be appropriately validated for their intended use.

For multi-story and complex reinforced concrete structures, the NRC has accepted the use of analytical codes intended for dynamic analysis. Determine if the use or absence of use of such codes is acceptable for specific analyses. The bases for acceptance may be the simplicity of the structure, extent or details provided in other calculations, or demonstration that the structural demands in the area of analysis are sufficiently low relative to the estimated capacity of the structure. Acceptance, on the basis of one or more of these conditions, should be such that further refinement of the computations would have negligible effect on the conclusion.

iii. Structural Evaluation

(1) Structural Capability

Review the selection or identification of the critical sections of the reinforced concrete structures to determine whether the structures conform with the design criteria regarding safety under the different load combinations. "Critical sections" are those that have the lowest margins of safety under the various loading conditions and types of stress. These sections may be selected on the basis of inspection, testing, sensitivity analysis, and/or finite-element analysis. The following paragraphs provide guidance for evaluating the identification of critical sections.

In particular, loads and stress demands for structures within the scope of ACI 359 (per CC-1100) shall be as defined and described in that standard.

Unless the lowest margins of safety have been determined by finite-element analysis using the applicable load combinations, critical sections should be identified for each structurally distinct element of the reinforced concrete structure. An integrally cast structure may have multiple structurally distinct elements (e.g., the different sides, base, and roof of a vault; and the base, corners, side walls, lips, and any structural discontinuities of an reinforced concrete cylinder such as at a trunnion).

The level of refinement needed in identifying critical sections depends primarily on the margins of safety and secondarily on the importance to safety.

Many reinforced concrete structures are designed primarily to provide radiation shielding. Such structures may have significantly excess capacity for structural loadings because of the use of section thicknesses selected for shielding and satisfying code requirements for minimum reinforcing steel. Structures important to safety may have such high margins of safety that only elementary structural computations are necessary to acceptably demonstrate compliance with all of the applicable load combinations. For simple elementary analysis, the margin of safety for a particular section should consider the highest axial, bending, and shear stresses occurring concurrently.

Intensive analysis is expected in order to prove that the truly critical sections are used when margins of safety are close to the minimum acceptable values.

The critical sections for bending, shear, axial stress, and combined stresses are typically different for a single structural element. They may also differ for different load combinations.

The lowest margins of safety for structural elements may result when different types of stresses exist under different load combinations.

Design and evaluation for accident loads involve structural loadings and responses that are not typically addressed in non-nuclear construction. In such instances, the structural shapes are not typical. Selection of representative sections for analysis by observation and experience may not be adequate without further computations to demonstrate that no other sections would have lower margins of safety. This could involve, for example, analyses of immediately adjacent sections to prove that margins of safety for the stress type increase in both directions from the section.

Table 3-1 identifies and describes loads used in combinations for reinforced concrete structures not within the scope of ACI 359. The symbols and terminology used in the SAR should correlate with these loads. However, if the symbols and terminology used in the SAR are different but acceptable, they may be used in the SER in place of those in Table 3-1, for consistency between the SER and SAR.

The NRC does not require analysis of load combinations for situations in which nuclear material is not present. However, reinforced concrete structures should not be exposed to credible damage that may not be evident or discovered before completion of construction or use. This could reduce the structural capacity or functional capability of the structure below that which is required. (For example, hidden damage could occur in handling and shipping precast reinforced concrete structures.)

Reinforced concrete structures subject to review but not "important to safety" should satisfy the load combinations of ACI 318, as a minimum.

(2) Fabrication and Construction

(a) Code Construction Criteria

Structures that are within the scope of ACI 359 must be fabricated and constructed in compliance with that standard. For reinforced concrete structures that are not within the scope of ACI 359, the NRC accepts construction in accordance with ACI 349 or ACI 318. Selection and validation of the proper concrete mix to meet design requirements is considered a construction function. By contrast, specification of cement type, aggregates, and special requirements for durability and elevated temperatures is considered a design or material selection function and is, therefore, governed by ACI 349 (and/or ACI 359, if applicable).

The following sections of ACI 318 (chapters, appendix, and paragraphing per ACI-318-95) have been accepted by the NRC for construction of ISFSI reinforced concrete structures that are not within the scope of ACI 359:

Chapter 1,	"General Requirements," Sections 1.1.1, 1.1.2, 1.1.3, and 1.1.5 (except references to design and material properties), and Section 1.3
Chapter 2,	"Definitions" (use ACI 349, Chapter 2)
Chapter 3,	"Materials," Sections 3.1 and 3.8 (except A-616, A-617, A-767, A-775, A-884, and A-934)
Chapter 4,	"Durability Requirements"
Chapter 5,	"Concrete Quality, Mixing, and Placing"
Chapter 6,	"Form Work, Embedded Pipes, and Construction Joints" (except references to design and material properties, which are governed by ACI 349)[b]

The following ASTM standard specifications also apply to construction and associated testing and are acceptable to the NRC:

C-31, C-33, C-39, C-42, C-94, C-109, C-150, C-172, C-192, C-260, C-494, C-496, C-685, and C-1017[c]

In addition, the following construction-related standards are identified in ACI 349 and may also be used:

C-88, C-131, C-289, and C-441

For reinforced concrete structures not important to safety, the NRC also accepts construction in accordance with ACI 318.

(b) Evaluation of Construction Commitments

Review the SAR documentation for inclusion of acceptable specifications for the planned construction and fabrication. Evaluate these specifications against the construction-related requirements in ACI 349 or ACI 318 for reinforced concrete structures not within the scope of ACI 359. For structures that are within the scope of ACI 359, the applicant must commit to fabricate and construct in accordance with ACI 359.

Construction requirements should prohibit the use of aluminum in any forms, chutes, ties, or other objects used in construction that will come in contact with wet concrete. This is because of the potential for concrete degradation as a result of an adverse chemical reaction with aluminum, including the fine particles that may be collected by wet concrete in pipes and chutes.

Construction specifications or drawing notes should preclude use of a greater amount, larger cross-section, or higher-yield of reinforcing steel than that derived by the design analysis. There is no constraint on use of higher strength concrete than that assumed by the design analysis if other properties required of the concrete mix are provided.

---

[b] Use ACI 349 for the remainder.

[c] C 1017 is only cited in ACI 318; all others are cited in both ACI 318 and ACI 349.

(3) Structural Compatibility with Functional Performance Requirements

Review the SAR documentation to ensure that the applicant's analysis of structural responses to DBE demonstrates that the SSC important to safety can continue to perform their intended safety functions. Note that the design codes and load combinations used for reinforced concrete structures important to safety can permit permanent deformation or other damage under design-basis event loading. If structural damage could occur, it is essential that the applicant demonstrate continuing capability with regard to essential functional performance. For reinforced concrete structures, this typically involves shielding the confinement cask from external events, maintaining cooling ventilation, and allowing ready retrieval of the confinement cask. Demonstrating continuing capability involves recognizing the nature and extent of credible damage that may occur and understanding potential interactions between the damaged reinforced concrete structures and other cask system structures important to safety.

Under conditions acceptable to the NRC, reinforced concrete structures are not required to survive an accident event with the same capability for a full design life and the same ability to withstand further accidents. Degradation of reinforced concrete structures should be readily apparent in the course of routine inspections and surveillances. Such degradation would also be discovered by the inspections and/or tests that may be proposed as responses to accidents. For example, tornado missile impact may degrade the radiation shielding by cratering a portion of the exterior of the reinforced concrete structure. Such degradation could be adequately repaired. The impact could also cause spalling at an inner, hidden surface, which could affect shielding and cooling air flow. The design should preclude the interior spalling unless the applicant proposes a practical means of detecting and remedying the situation.

The NRC has accepted returning the stored fuel to the spent fuel pool or transfer to an undamaged cask and not making further use of the damaged component, as remedial action for an accident-level event. The NRC may also accept inspection and repair of structural damage, depending on the design and proposed actions and the feasibility of both detection and repair.

3. Other System Components Important to Safety

a. Scope

Subsections V.1.d (i), and (ii), above, provide general guidance for the structural review of cask system components. This portion of the DCSS structural review supplements that guidance by addressing procedures for evaluating all structures that are important to safety (as defined in 10 CFR Part 72), but are not addressed as components of the confinement cask (Subsection V.1, above) and are not constructed using reinforced concrete (Subsection V.2, above). Structures may include items such as gamma and neutron shielding, overpack material, and any respective encasement. This evaluation should include drawings, plans, sections, and technical specifications for these SSC.

b. Structural Design Criteria and Design Features

   i. Design Criteria

   (1) General Structural Requirements

Structural requirements are driven by the functional roles of the system components and the need to maintain safety. Safety requirements are expressed in the referenced rules, standards, and codes and as criteria specific to the component. The basic safety requirements are that the structural and functional design must preclude the following:

- unacceptable risk of criticality

- unacceptable release of radioactive materials to the environment

- unacceptable radiation dose to the public or workers

- significant impairment of ready retrievability of stored nuclear materials

(2) Applicable Codes and Standards

The NRC accepts the use of ANSI/ANS-57.9 (together with the codes and standards cited therein) as the basic reference for ISFSI structures important to safety that are not designed in accordance with the Section III of the ASME B&PV Code. However, both the lifting equipment design and the devices for lifting system components that are important to safety must comply with ANSI Standard N14.6.

The NRC accepts the load combinations shown in Table 3-1 for structures not designed under either Section III of the ASME B&PV Code or ACI 359. These load combinations are defined on the basis of ANSI/ANS-57.9, with supplemental definition of terms and combinations.

Review the suitability of the applicant's identification of codes and standards that are to be met by the structural design and construction of other components subject to NRC approval. The principal codes and standards include the following references that may apply to steel structures and components:

- AISC, "Specification for Structural Steel Buildings — Allowable Stress Design and Plastic Design"

The NRC has not yet received any applications that propose a steel design on the basis of the AISC's "Load and Resistance Factor Design (LRFD) Specification for Structural Steel Buildings." If such a design was received, the NRC would evaluate the proposal for compliance with the load combinations summarized in Table 3-1 and for consistent application of the LRFD design methodology.

- AWS D1.1, "Structural Welding Code Steel"

- ASCE 7 "Minimum Design Loads for Buildings and Other Structures" [however, note that load combinations established on the basis of ANSI/ANS-57.9 (DCSS SRP Table 3-1) are to be used]

- ACI 349, Appendix B, for embedments or 10.14 for composite compression sections, as applicable, when constructed of structural steel embedded in reinforced concrete [Where requirements do not conflict, the steel must also comply with the requirements of the codes stated above. In addition, ACI 349 defines constraints for obtaining ductile response to extreme loads by ensuring that the strength of steel embedments controls the design; these constraints must not be subverted by overdesign of the steel.]

These documents cite further sources of criteria, which are considered to have the effect of being directly cited or quoted in the basic structural criteria. In addition, the NRC accepts ANSI N14.6 as the basis for design of lifting equipment and components of vessels and other devices provided for lifting, where such equipment is important to safety.

To date, the NRC has not required applicants to design or build steel ISFSI structures important to safety in compliance with ANSI/ANS N690, "Nuclear Facilities — Steel Safety-Related Structures for Design Fabrication and Erection"[46].

For fluid systems that may be connected to a penetration of the confinement barrier outside an enclosing structure licensed under 10 CFR Part 50 (e.g., the fuel pool building), the NRC accepts construction consistent with requirements comparable to those used for Quality Group C, as shown in RG 1.26[47] and NUREG-0800[48], Section 3.2.2. (In this context, "construction" includes materials, design, fabrication, examination, testing, inspection, and certification required in the manufacture and installation of components.) If analysis shows that the maximum conservatively estimated offsite dose would not exceed 0.5 rem to the whole body or any equivalent part of the body, the NRC may accept construction that satisfies Quality Group D. (In this instance, the NRC accepts the analysis procedure identified in RG 1.262, Subsection C.2.d.)

Quality Group C requires construction of piping, pumps, valves, atmospheric storage tanks, and 0–15 psig storage tanks in conformance with Section III of ASME B&PV Code 1, Class 3 (Subsection ND). In addition, Quality Group C requires that supports for these components meet the requirements of Subsection NF.

By contrast, Quality Group D requires compliance with the following codes, as a minimum:

Piping: ANSI/ASME B31.1, "Power Piping"[49]

Pumps: Manufacturer's standards

Valves: ANSI/ASME B31.1 and ANSI B16.34[50]

Atmospheric
storage tanks: AWWA D100, "Standard for Steel Tanks — Standpipes, Reservoirs, and Elevated Tanks for Water Storage"[51] or ANSI/ASME B96.1, "Specification for Welded Aluminum-Alloy Field-Erected Storage Tanks"[52]

0–15 psig
storage tanks: API 620, "Recommended Rules for Design and Construction of Large, Welded, Low-Pressure Storage Tanks"[53]

The NRC accepts the "Boundaries of Jurisdiction" applicable to Section III, Subsections NB-1130 and NC-1130, of ASME B&PV Code 1. These boundaries apply to attachments to penetrations of the confinement barrier outside an enclosure licensed under 10 CFR Part 50. Specifically, these boundaries define whether the attachments must be designed, fabricated, and installed in accordance with Section III, Subsection NB or NC, of ASME B&PV Code 1.

The NRC has not yet received any applications for licensing or approval of a cask system that included masonry important to safety. Masonry is not considered suitable for confinement, but it may be acceptable for enclosures and physical or radiation shielding applications.

In evaluating any future applications that may contain masonry important to safety or otherwise subject to NRC approval, reviewers should focus on the following assessments:

- Determine the acceptability of codes, standards, and specifications used for design, materials, and construction.

- Review the use of and need for any supplementary design or construction criteria appropriate to the analysis or application.

- Evaluate compliance with the codes, standards, specifications, and supplementary criteria. Use of the strength design of masonry procedures (as defined in the Uniform Building Code 32 and ACI 530[54]) would provide the closest parallel to the procedures accepted for reinforced concrete design.

- Evaluate load combinations used for structural analysis and determination of safety margins. These load combinations for masonry structures should include the loads identified in Table 3-1. Specifically, the applicant should select the more conservative load combinations from either those identified in Table 3-1 for reinforced concrete strength design and those specified in the selected masonry code. The appropriate strength reduction factors should also be applied.

The evaluation process may also involve confirmatory staff computations. Specifically, these computations should verify the submitted margins of safety and confirm that the applicant has identified the lowest margins of safety for the significant structural elements.

     ii.   Structural Design Features

Review the design description in the SAR documentation to ensure that it defines the functional performance required of the structures. The design description should provide for the corresponding capability.

Auxiliary cask system equipment important to safety has often been specially designed. In particular, the structural design features that provide for safety should be supported by design or operational analysis. This analysis should demonstrate that the equipment will meet the basic safety criteria, regardless of problems that may occur in mechanical, electrical, human operator, or other operations.

The NRC has accepted and approved cask system designs that depend on the operation of new mechanical systems for system use. NRC approval does not certify that the mechanical systems will operate as projected, but rather, that proper functioning is necessary to successfully complete a specified operation. Such approval reflects a finding by the NRC staff that, regardless of the system's success (or lack thereof) in mechanical operation, the basic safety criteria will be met, as stated above.

In accordance with 10 CFR 72.24(i) the SARshall include a schedule showing how the applicant will resolve any safety questions regarding functional adequacy or reliability. The review should focus on resolution of all safety questions before license approval. Safety questions that cannot feasiblely be resolved before license approval should be included in the conditions of use (or license conditions) in the SER.

Review the proposed system design against planned normal, and off-normal, operations and accidents. Determine whether the structural design of the equipment provides for continuing satisfaction of the basic safety criteria. Consider that the equipment could fail to operate at any time (i.e., during operations at the physical limits of speed or range, or during a credible, off-normal, or accident event).

c. Structural Materials

The SAR documentation should fully define the structural materials used for components important to safety that are not addressed in Subsections V.1 and V.2, above. In addition, the SAR should identify properties related to structural performance and resistance or response to thermal, radiation, or other applicable environments.

Confirm that these materials and their properties are derived from acceptable sources. Ensure that the SAR addresses resistance to corrosion in the prospective environments, including the need for protective coatings, protective coating renewal, and/or corrosion allowances. In addition, the SAR should cite controls imposed on material quality, or such controls should be included in codes that are incorporated by reference in the design, fabrication, and construction criteria. The reviewer should be familiar with the information contained within NRC bulletin 96-04: Chemical, Galvanic or other Reactions in Spent fuel Storage and Transportation Casks.

d. Structural Analysis

Subsections V.1.d(i) and (ii) provide guidance regarding structural analysis for cask system structures in general. These subsections provides supplemental guidance primarily related to steel structures, other than the confinement cask and its contents and integral components, which are typically designed to Section III of the ASME B&PV Code

    i. Load Conditions

The load definitions and combinations shown in Table 3-1 have been accepted by the NRC for analysis of steel and reinforced concrete ISFSI structures that are important to safety. These load combinations are included in or derived from ANSI/ANS 57.9 and ACI 349.

Structures that are important to safety should have sufficient capability for every section to withstand the worst-case loads under normal and off-normal conditions. Such capability ensures that these structures will not experience permanent deformation or degradation of the capability to withstand any future loadings.

The NRC accepts the load combinations in Table 3-1, which implement and supplement those of ANSI/ANS-57.9.

    ii. Structural Analysis Methods

The applicant should select and use analytical methods that are appropriate for the proposed type of materials and construction. In certain instances, however, the applicant may have to adapt existing analytical methods, codes, and models for highly specialized cask system equipment designs. Such instances require special review attention. In particular, ensure that the adapted approach is fully documented, supported, and acceptable. In addition, consider the potential for safety-related risk associated with a possible error in the design of special cask system equipment. The degree of risk indicates the suitability and acceptability of the adapted approach.

iii. Structural Evaluation

In evaluating the variety of cask system equipment and structures that may be important to safety, ensure compliance with the basic safety criteria in Subsection V.3.b(i)(1), above.

4. Other Components Subject to NRC Approval

a. Scope

The cask system description provided in the SAR may include a variety of structures that are not important to safety, such as transporters, ram systems vacuum drying systems, and drain and fill quick disconnects. Review these structures to ensure proper functioning to the extent that the structures represent required elements of the total cask system. In particular, evaluate all structures that are proposed for approval in a design acceptable to the NRC. This evaluation should ensure that the SAR provides sufficient information to confirm the proper functioning of the structures and the overall system. For each system element that is not important to safety, address the potential response to accidents and natural phenomenon events, in order to ensure that the given element will not jeopardize the safety provided by other system elements.

In addition, to the extent that physical protection is incorporated in the cask system design subject to approval, reviewed the design for compliance with 10 CFR 72.182.

b. Structural Design Criteria and Design Features

i. Design Criteria

(1) General Structural Requirements

Structures subject to approval but not important to safety should be reviewed on the based of determining whether the structures can properly perform their intended function(s). In addition, the NRC review should ensure that the response of the structures to credible off-normal and accidents and conditions will not create secondary hazards for cask system components or the stored nuclear materials.

(2) Applicable Codes and Standards

Review the suitability of the applicant's identification of codes and standards to be met by the structural design and construction of other components subject to NRC approval. The principal codes and standards include the following references:

- ASCE 7

- Uniform Building Code (UBC)

- AISC, "Specification for Structural Steel Buildings—Allowable Stress Design and Plastic Design"

- AISC "Code of Standard Practice for Steel Buildings and Bridges"

- ASME B&PV Code, Section VIII

ii. Structural Design Features

Review the adequacy of the applicant's descriptions of cask system components that are not important to safety, but are subject to NRC approval. These descriptions should adequately identify the intended function(s) of each component.

Although the components evaluated in this portion of the DCSS review are not directly important to safety, a credible possibility may exist that the structural response or failure of these components may cause a secondary risk to other components that *are* important to safety or to the subject nuclear material. For example, under tornado or seismic event conditions, the components may impact other components that are important to safety. When such a possibility exists, the applicant must provide more extensive structural information and greater assurance of acceptable fabrication and construction.

c. Structural Materials

Review the identification of structural materials to the extent appropriate to determine if they are adequate for their intended function(s). Determine the required level of review and extent of information in relation to the possibility and consequences of secondary effects on components that are important to safety (see b(ii), above). Materials should be as permitted or specified in the applicable code(s) (see V.3 b(i), above).

d. Structural Analysis

i. Load Conditions

The load definitions and combinations shown in Table 3-1 have been accepted by the NRC for analysis of steel and reinforced concrete ISFSI structures that are important to safety. These load combinations may also be used for structures not important to safety.

In addition, for structures not important to safety, the NRC accepts the use of load combinations given in the Uniform Building Code, as well as ACI 349, ANSI/ANS 57.9, and ASCE 7.

The NRC also accepts the load descriptions, combinations, and analytical approaches given in the ASME B&PV Code, Section VIII, for pressure systems, vessels, and casks that do not form elements of the confinement cask.

ii. Structural Analysis Methods

The reviewer shall evaluate the applicant's selection and use of structural analysis methods, codes, and models and ensure that these are consistent with and appropriate for the design code applicable to the component (as discussed above).

iii. Structural Evaluation

The reviewer may determine that an NRC structural evaluation of certain other components is not necessary for approval of the cask system. Similarly, the NRC may determine that approval of the cask system does not need to include specific components that are not important to safety, even though the applicant seeks approval of those components as part of the application.

The SER should identify the system components that are excluded from the approval, stating the rationale for exclusion of each. As a corollary, the SER should also identify the components that are included, stating any limitations on the scope of the NRC review (e.g., "reviewed for functionality only").

VI. Evaluation Findings

The structural evaluation must provide reasonable assurance that the cask system will allow safe storage of spent fuel. This finding should be reached on the basis of a review that considered the regulation, appropriate Regulatory Guides, applicable codes and standards, and accepted engineering practices. Acceptance of the structural design of a storage cask system therefore implies that the design meets the relevant requirements of the following regulations:

- The SAR adequately describes all structures, systems, and components (SSC) that are important to safety, providing drawings and text in sufficient detail to allow evaluation of their structural effectiveness.

- The applicant has met the requirements of 10 CFR 72.24, "Contents of Application: Technical Information," with regard to information pertinent to structural evaluation.

- The applicant has met the requirements of 10 CFR 72.26, "Contents of Application," and 10 CFR 72.44(c), "License Conditions," with regard to technical specifications pertaining to the structures of the proposed cask system.

- The applicant has met the requirements of 10 CFR Part 72.122(b) and (c) and 10 CFR Part 72.24(c)(3). The structures, systems, and components important to safety are designed to accommodate the combined loads of normal, off-normal, accident, and natural phenomena events with an adequate margin of safety. Stresses at various locations of the cask for various design loads are determined by analysis. Total stresses for the combined loads of normal, off normal, accident, and natural phenomena events are acceptable and are found to be within limits of applicable codes, standards, and specifications.

- The applicant has met the requirements of 10 CFR Part 72.124(a), "Criteria for Nuclear Criticality Safety", and 10 CFR Part 72.236 (b), "Specific requirements for spent fuel storage cask approval." The structural design and fabrication of the DCSS includes structural margins of safety for those SSC important to nuclear criticality safety. The applicant has demonstrated adequate structural safety for the handling, packaging, transfer, and storage under normal, off-normal, and accident conditions.

- The applicant has met the requirements of 10 CFR 72.236(l), "Specific Requirements for Spent Fuel Storage Cask Approval." The design analysis and submitted bases for evaluation acceptably demonstrate that the cask and other systems important to safety will reasonably maintain confinement of radioactive material under normal, off-normal, and credible accident conditions.

- The applicant has met the requirements of 10 CFR 72.120, "General Considerations," and 10 CFR 72.122, "Overall Requirements," with regard to inclusion of the following provisions in the structural design:

    - design, fabrication, erection, and testing to acceptable quality standards

    - adequate structural protection against environmental conditions and natural phenomena, fires, and explosions

    - appropriate inspection, maintenance, and testing

    - adequate accessibility in emergencies

    - a confinement barrier that acceptably protects the spent fuel cladding during storage

    - structures that are compatible with appropriate monitoring systems

    - structural designs that are compatible with ready retrievability of spent fuel

- The applicant has met the specific requirements of 10 CFR 72.236(e)(f) (g)(h)(i)(j)(k) and (m), as they apply to the structural design for spent fuel storage cask approval. The cask system structural design acceptably provides for the following required provisions:

    - redundant sealing of confinement systems

    - adequate heat removal without active cooling systems

    - storage of the spent fuel for a minimum of 20 years

    - compatibility with wet or dry spent fuel loading and unloading facilities

    - acceptable ease of decontamination

    - inspections for defects that might reduce confinement effectiveness

    - conspicuous and durable marking

    - compatibility with removal of the stored fuel from the site, transportation, and ultimate disposition by the U.S. Department of Energy

Table 3-1 Loads and load combinations

Designations and Descriptions of Loads

Definitions of terms used in the following table are as accepted by the NRC. Many definitions are expanded, with their intended applications more fully described and implemented than in the referenced sources.

Table 3-1 does not apply to the analysis of confinement casks and other components designed in accordance with Section III of the ASME B&PV Code.

Capacities ("S" and "U" terms) and demands (factored or unfactored loads) may be loads, forces, moments, or stresses caused by such loads. Usage must be consistent among the terms used in the load combination. Units of force, rather than mass, are to be used for loads.

Definitions of terms used in the load combination expressions for reinforced concrete and steel are derived from ANSI 57.9, ACI 349, AISC specifications, or another source. Where used in an expression related to steel analysis, definitions derived from ACI 349 are not limited in application to reinforced concrete analyses.

The load combinations defined on the basis of allowable stress apply to total stresses (that is, combined primary and secondary stresses). The load and stress factors do not change if secondary stresses are included.

Symbol	Capacity or Load Term	Source	Capacity or Load (or Demand) Description
S	Steel ASD strength	ANSI 57.9	Strength of a steel section, member, or connection computed in accordance with the "allowable stress method" of the AISC "Specification for Structural Steel Buildings." Note exception for shear strength (see S_v).
S_v	Steel ASD shear strength	ANSI 57.9	Shear strength of a section, member, or connection computed in accordance with the "allowable stress method" of the AISC "Specification for Structural Steel Buildings."
Us	Steel plastic strength	ANSI 57.9	Strength (capacity) of a steel section, member, or connection computed in accordance with the "plastic strength method" of the AISC "Specification for Structural Steel Buildings."
U_c	reinforced concrete available strength	ANSI 57.9, ACI 349	Minimum available strength (capacity) of reinforced concrete section, member, or embedment to meet the load combination, calculated in accordance with the requirements and assumptions of ACI 349 and, after application of the strength reduction factor, $\emptyset$, as defined and prescribed at §9.2, "Design Strength," of ACI 349. If strength may be reduced during the design life by differential settlement, creep, or shrinkage, those effects shall be incorporated in the dead load, D (instead of by subtraction from minimum available strength). reinforced concrete footing and foundation sections whose demand loads are dominated by the maximum soil reaction may be designed and evaluated using U_f.

Symbol	Capacity or Load Term	Source	Capacity or Load (or Demand) Description
U_f	Strength of foundation sections	ANSI 57.9	Minimum available strength of reinforced concrete footing and foundation sections whose demand loads are dominated by the maximum soil reaction, and after the strength reduction factor, $\emptyset$, as defined and prescribed at §9.2, "Design Strength," of ACI 349 is applied. Structural elements interface with columns, walls, grade beams, or footings and foundations should be evaluated by using load factors and load combinations for U_c. These interface elements include anchor bolts and other embedments, dowels, lugs, keys, and reinforcing extended into the footing or foundation.
U_g	Soil reaction or pile capacity	ANSI 57.9	Minimum available soil reaction or pile capacity is determined by foundation analysis (included with a license application SAR, or expressed in a SAR for approval of a cask system as a required minimum for the cask system design). U_g is derived using the same load factors and load combinations as shown for determination of Uc.
O/S	Overturning/ sliding resistance	ANSI 57.9	Required minimum available resistance capacity of structural unit against both overturning or sliding. Capacities for resistance of overturning and sliding are checked against the factored load combination separately, although the minimum margins of safety may occur concurrently. O/S is not determined by strength capacities of structural elements. Stress or strength demands resulting from an overturning or sliding situation are evaluated in load combinations involving S, S_v, U_s, U_c, and U_f.
	All loads used in combination	ANSI 57.9 ACI 349	If any load reduces the effects of the combination of the other loads, and that load would always be present in the condition of the specific load combination, the net coefficient (factor) for that load shall be taken as 0.90. If the load may not always be present, the coefficient for that load shall be taken as zero. Each load that may not always be present in the load combinations is to be varied from 0 to 100% to simulate the most adverse loading conditions (to the extent of proving that the lowest margins of safety have been determined).

Symbol	Capacity or Load Term	Source	Capacity or Load (or Demand) Description
D	Dead load	ANSI 57.9	Dead load of the structure and attachments including permanently installed equipment and piping. The weight and static pressure of stored fluids may be included as dead loads when these are accurately known or enveloped by conservative estimates. Loads resulting from differential settlement, creep, and/or shrinkage, if they produce the most adverse loading conditions, are included in dead load. If differential settlement, creep, or shrinkage would reduce the combined loads, it shall be neglected. D includes the weight of soil vertically over a footing or foundation for the purposes of determining U_g, U_f and O/S. Regardless of the load combination factor applied, D is to be varied by +5% if that produces the most adverse loading condition.
L	Live loads	ANSI 57.9 ACI 349	Live loads, including equipment (such as a loaded storage cask) and piping not permanently installed, and all loads other than dead loads that might be experienced that are not separately identified and used in the load combination, and that are applicable to the situation addressed by the load combination. Typically includes the gravity and operational loads associated with handling equipment, and routine snow, rain, ice, and wind loads and normal and off-normal impacts of equipment. Loads attributable to piping and equipment reactions are included. Depending on the case being analyzed, may include normal or off-normal events not separately identified, as may be caused by handling (not including drop), equipment or instrument malfunction, negligence, and other man-made or natural causes. Live loads attributable to casks with stored fuel need only be varied by credible increments of loading of an individual cask. Live loads attributable to multiple casks should be varied for the presence and positioning of one or more cask(s), as necessary and varied to determine the lowest margins of safety.

Symbol	Capacity or Load Term	Source	Capacity or Load (or Demand) Description
L	Live load for precast structures before final integration in-place		Live loads for precast structures shall consider all loading and restraint conditions from initial fabrication to completion of the structure, including form removal, storage, transportation, and erection. The NRC is only concerned with analysis of loading of reinforced concrete structures before use for ISFSI functions to the extent that the structures should not risk damage that may not be evident, thereby jeopardizing the capacity of the structures when in use. If the damage would be visibly obvious before installation, analysis of capacity versus pre-completion demands is not required.
DB_	"Design-basis" (accident-level) loads	10 CFR 72	Design-basis loads are controlling bounds for the following external event estimates: (1) extreme credible natural events to be used for deriving design bases that consider historical data or rated parameters, physical data, or analysis of upper limits of the physical processes involved (2) extreme credible external man-induced events used for deriving design bases on the basis of analysis of human activity in the region taking into account the site characteristics and associated risks. Design-basis loads include credible accidents and extreme natural phenomena. Presumption of concurrent independent accidents or severe natural phenomena producing compounding design-basis loads is not required. Capacity to resist design basis loads can be assumed to be that of a structure that has not been degraded by previous design basis loads, unless prior significant degradation in structural capacity may credibly occur and remain undetected.
T	Thermal loads	ANSI 57.9 NUREG-0800, Section 3.8.3)	Thermal loads, including loads associated with "normal" condition temperatures, temperature distributions, and thermal gradients within the structure; expansions and contractions of components; and restraints to expansions and contractions, with the exception of thermal loads that are separately identified and used in the load combination. Thermal loads shall presume that all loaded fuel has the maximum thermal output allowed at time of initial loading in the ISFSI. Thermal loads shall be determined for the most severe of both steady-state and transient conditions. For multiple cask storage facilities, thermal loads shall be determined for the worst-case loadings on potentially critical sections (e.g., all in place, only one cask in place, alternating casks in place).

Symbol	Capacity or Load Term	Source	Capacity or Load (or Demand) Description
T_a	Accident-level thermal loads	ANSI 57.9, ACI 349	Thermal loads produced directly or as a result of *off-normal or design-basis* accidents, fires, or natural phenomena. [Note: Although off-normal and design-basis thermal loads are treated the same in the load combinations, there is a distinction between off-normal and design-basis temperature limits for concrete. Off-normal temperature limits are the same as for "normal" conditions.] For multiple cask storage facilities, thermal loads shall be determined for the worst-case loadings on potentially critical sections.
A	Accident loads	ANSI 57.9	Loads attributable to the direct and secondary effects of an off-normal or design-basis accident, as could result from an explosion, crash, drop, impact, collapse, gross negligence, or other man-induced occurrences; or from severe natural phenomena not separately defined. Loads attributable to direct and secondary effects may be assumed to be non-concurrent unless they might be additive. The capacity for resistance to the demand resulting from secondary effects would be that residual capacity following any degradation caused by the direct effect.
H	Lateral soil pressure	ANSI 57.9	Loads caused by lateral soil pressure as would exist in normal, off-normal, or design-basis conditions corresponding to the load combination in which used. H includes lateral pressure resulting from ground water, the weight of the earth, and loads external to the structure transmitted to the structure by lateral earth pressure (not including earthquake loads, which are included in E, see below). H does not include soil reaction associated with attempted lateral movement of the structure or structural element in contact with the earth.
G	Loads attributable to soil reaction	ANSI 57.9	Used only in load combinations for footing and foundation structural sections for which demand is limited by the soil reactions. G represents loads attributable to the maximum soil reaction (horizontal (passive pressure limit) and vertical (soil or pile bearing limit)) that would exist in normal, off-normal, or design-basis conditions, corresponding to the load combination in which used. G is a function of U_g (i.e., $G = f(U_g)$).
W	Wind loads	ACI 349	Winds loads produced by normal and off-normal maximum winds. Pressure resulting from wind and with consideration of wind velocity, structure configuration, location, height above ground, gusting, importance to safety, and elevation may be calculated as provided by ASCE 7.

Symbol	Capacity or Load Term	Source	Capacity or Load (or Demand) Description
W_t	Tornado loads	ACI 349	Loads attributable to wind pressure and wind-generated missiles caused by the design-basis tornado or design-basis wind (for sites where design-basis wind rather than tornado produces the most severe pressure and missile loads). Pressure resulting from wind velocity and elevation may be calculated as provided for these factors in ASCE 7. Tornado wind velocity or pressure does not have to be increased for structure importance, gusting, location, height above ground, or importance to safety (these do apply for design-basis wind).
E	Earthquake loads		Loads attributable to the direct and secondary effects of the design-basis or off-normal flood, including flooding caused by severe and extreme natural phenomena (e.g., seiches, tsunamis, storm surges), dam failure, fire suppression, and other accidents.

Load Combinations for Steel and reinforced Concrete (reinforced concrete) Structures

The reinforced concrete structure load combinations apply to reinforced concrete structures important to safety that are not within the scope of ACI 359 (ASME B&PV Code, Section III, Division 2). The load combinations apply to steel structures important to safety that are not within the scope of the ASME B&PV Code, Section III, Division 1. The NRC accepts, but does not require use of these load combinations for steel and reinforced concrete structures that are not important to safety. The NRC accepts steel analyses that reflect allowable stress design or plastic strength design. Steel load combinations may be determined on the basis of the set of load combination expressions involving either "S" or "U$_s$."

Load Combination	Derivation (Reference)	Acceptance Criteria
Reinforced Concrete Structures — Normal Events and Conditions		
$U_c > 1.4\,D + 1.7\,L$	ANSI 57.9	Capacity/demand >1.00 for all sections
$U_c > 1.4\,D + 1.7\,(L + H)$	ANSI 57.9	Capacity/demand >1.00 for all sections
Reinforced Concrete Structures — Off-Normal Events and Conditions		
$U_c > 1.05\,D + 1.275\,(L + H + T)$	ANSI 57.9	Capacity/demand >1.00 for all sections
$U_c > 1.05\,D + 1.275\,(L + H + T + W)$	ANSI 57.9	Capacity/demand >1.00 for all sections

Load Combination	Derivation (Reference)	Acceptance Criteria

Reinforced Concrete Structures — Accidents and Conditions

Load Combination	Derivation (Reference)	Acceptance Criteria
$U_c > D + L + H + T + (E \text{ or } F)$	ANSI 57.9	Capacity/demand >1.00 for all sections.
$U_c > D + L + H + T + A$	ANSI 57.9	Capacity/demand >1.00 for all sections. An overturning accident for a cask in transfer or in separate storage on a pad is to be assumed, unless more severe overturning also occurs as a result of a natural phenomenon.
$U_c > D + L + H + T_a$	ANSI 57.9	Capacity/demand >1.00 for all sections.
$U_c > D + L + H + T + W_t$	ANSI 57.9 ACI 349	The load combination (capacity/demand >1.00 for all sections) shall be satisfied without missile loadings. Missile loadings are additive (concurrent) to the loads caused by the wind pressure and other loads; however, local damage may be permitted at the area of impact if there will be no loss of intended function of any structure important to safety.

Reinforced Concrete Footings/Foundations — Normal Events and Conditions

Load Combination	Derivation (Reference)	Acceptance Criteria
$U_f > D + (L + G)$	ANSI 57.9 ACI 349	Capacity/demand >1.00 for all sections. For footing and foundation sections with load limited by soil reaction.
$U_f > D + (L + H + G)$	ANSI 57.9 ACI 349	Capacity/demand >1.00 for all sections. For footing and foundation sections with load limited by soil reaction.

Reinforced Concrete Footings/Foundations — Off-Normal Events and Conditions

Load Combination	Derivation (Reference)	Acceptance Criteria
$U_f > D + (L + H + T + G)$	ANSI 57.9 ACI 349	Capacity/demand >1.00 for all sections. For footing and foundation sections with load limited by soil reaction
$U_f > D + (L + H + T + W + G)$	ANSI 57.9 ACI 349	Capacity/demand >1.00 for all sections. For footing and foundation sections with load limited by soil reaction.

Reinforced Concrete Footings/Foundations — Accident-Level Events and Conditions

Load Combination	Derivation (Reference)	Acceptance Criteria
$U_f > D + L + H + T + E + G$	ANSI 57.9 ACI 349	Capacity/demand >1.00 for all sections. For footing and foundation sections with load limited by soil reaction.

Structural Evaluation

Load Combination	Derivation (Reference)	Acceptance Criteria
$U_f > D + L + H + T + A + G$	ANSI 57.9 ACI 349	Capacity/demand >1.00 for all sections. For footing and foundation sections with load limited by soil reaction.
$U_f > D + L + H + T_a + G$	ANSI 57.9 ACI 349	Capacity/demand >1.00 for all sections. For footing and foundation sections with load limited by soil reaction.
$U_f > D + L + H + T + W_t + G$	ANSI 57.9 ACI 349	Capacity/demand >1.00 for all sections. For footing and foundation sections with load limited by soil reaction.
$U_f > D + L + H + T + F + G$	ANSI 57.9 ACI 349	Capacity/demand >1.00 for all sections. For footing and foundation sections with load limited by soil reaction.

Steel Structures Allowable Stress Design — Normal Events and Conditions

$(S \text{ and } S_v) > D + L$	ANSI 57.9	Factored strength/demand >1.00 for all sections
$(S \text{ and } S_v) > D + L + H$	ANSI 57.9	Factored strength /demand >1.00 for all sections

Steel Structures Allowable Stress Design — Off-Normal Events and Conditions

$1.3\,(S \text{ and } S_v) > D + L + H + W$	ANSI 57.9	Factored strength /demand >1.00 for all sections
$1.5\,S > D + L + H + T + W$	ANSI 57.9	Factored strength/demand >1.00 for all sections. Thermal loads may be neglected when analysis shows that they are secondary and self-limiting in nature, and when the material is ductile.
$1.4\,S_v > D + L + H + T + W$	ANSI 57.9	Factored strength/demand >1.00 for all sections. Thermal loads may be neglected when analysis shows that they are secondary and self-limiting in nature, and when the material is ductile.

Load Combination	Derivation (Reference)	Acceptance Criteria
Steel Structures Allowable Stress Design — accidents and Conditions		
$1.6\ S > D + L + H + T +$ (E or W_t or F)	ANSI 57.9; ACI 349	Factored strength/demand >1.00 for all sections. Thermal loads may be neglected when analysis shows that they are secondary and self-limiting in nature, and when the material is ductile.
$1.4\ S_v > D + L + H + T +$ (E or W_t or F)	ANSI 57.9; ACI 349	Factored strength (allowable stress design)/demand >1.00 for all sections. Thermal loads may be neglected when analysis shows that they are secondary and self-limiting in nature, and when the material is ductile.
$1.7\ S > D + L + H + T + A$	ANSI 57.9	Factored strength/demand >1.00 for all sections. Thermal loads may be neglected when analysis shows that they are secondary and self-limiting in nature, and when the material is ductile..
$1.4\ S_v > D + L + H + T + A$	ANSI 57.9	Factored strength/demand >1.00 for all sections. Thermal loads may be neglected when analysis shows that they are secondary and self-limiting in nature, and when the material is ductile.
$1.7\ S > D + L + H + T_a$	ANSI 57.9	Factored strength/demand >1.00 for all sections
$1.4\ S_v > D + L + H + T_a$	ANSI 57.9	Factored strength/demand >1.00 for all sections
Steel Structures Plastic Strength Design — Normal Events and Conditions		
$U_s > 1.7\ (D + L)$	ANSI 57.9	Plastic capacity/ demand >1.00 for all sections
$U_s > 1.7\ (D + L + H)$	ANSI 57.9	Plastic capacity/ demand >1.00 for all sections
Steel Structures Plastic Strength Design — Off-Normal Events and Conditions		
$U_s > 1.3\ (D + L + H + W)$	ANSI 57.9	Plastic capacity/ demand >1.00 for all sections.

Load Combination	Derivation (Reference)	Acceptance Criteria
$U_s > 1.3 (D + L + H + T + W)$	ANSI 57.9	Plastic capacity/ demand >1.00 for all sections. Thermal loads may be neglected when analysis shows that they are secondary and self-limiting in nature, and when the material is ductile.

Steel Structures Plastic Strength Design — accidents and Conditions

Load Combination	Derivation (Reference)	Acceptance Criteria
$U_s > 1.1 (D + L + H + T + (E \text{ or } W_t \text{ or } F))$	ANSI 57.9	Plastic capacity/ demand >1.00 for all sections. Thermal loads may be neglected when analysis shows that they are secondary and self-limiting in nature, and when the material is ductile. The load combination (capacity/demand >1.00 for all sections) shall be satisfied without missile loadings. Missile loadings are additive (concurrent) to the loads caused by the wind pressure and other loads, however local damage may be permitted at the area of impact if there will be no loss of intended function of any structure important to safety.
$U_s > 1.1 (D + L + H + T + A)$	ANSI 57.9	Plastic capacity/ demand >1.00 for all sections. An overturning accident for a cask in transfer or in separate storage on a pad is to be assumed unless more severe overturning also occurs as a result of a natural phenomenon. Thermal loads may be neglected when analysis shows that they are secondary and self-limiting in nature, and when the material is ductile.
$U_s > 1.1 (D + L + H + T_a)$	ANSI 57.9	Plastic capacity/ demand >1.00 for all sections

Overturning and Sliding — Normal and Off-Normal Events and Conditions

Load Combination	Derivation (Reference)	Acceptance Criteria
$O/S \geq 1.5 (D + H)$	ANSI 57.9	Capacity/demand ≥1.00 for structure, to be satisfied for both overturning and sliding

Overturning and Sliding — accidents and Conditions

Load Combination	Derivation (Reference)	Acceptance Criteria
$O/S \geq 1.1 (D + H + E)$	ANSI 57.9	Capacity/demand ≥1.00 for structure, to be satisfied for both overturning and sliding

Load Combination	Derivation (Reference)	Acceptance Criteria
$O/S \geq 1.1\,(D + H + W_t)$	ANSI 57.9	Capacity/demand ≥ 1.00 for structure, to be satisfied for both overturning and sliding

VII. References

Except for Federal regulations, the documents listed below are suitable for use as references in SARs relevant to structural design and evaluation to the extent described in this chapter. The citations below refer to the latest version of each document, except where a specific edition is indicated. References noted in the documents cited below are considered incorporated to this list. References to "Parts" of the *U.S. Code of Federal Regulations* shall be presumed to imply the "current Code," which includes all changes effective as of the date of submission of the application for approval.

1. *U.S. Code of Federal Regulations*, Part 72, "Licensing Requirements for the Independent Storage of Spent Nuclear Fuel and High-Level Radioactive Waste," Title 10, "Energy"

2. *U.S. Code of Federal Regulations*, Part 100, "Reactor Site Criteria' Title 10, "Energy"

3. U.S. Nuclear Regulatory Commission, "System Evaluation Program", Office of Nuclear Reactor Regulation

4. American Society of Mechanical Engineers (ASME) Boiler and Pressure Vessel (B&PV) Code, Section III, "Rules for Construction of Nuclear Power Plant Components"

5. American National Standards Institute, American Nuclear Society, " Design Criteria for an Independent Spent Fuel Storage Installation (Dry Storage Type)," ANSI/AND-57.9

6. American Society of Mechanical Engineers, Boiler and Pressure Vessel Code, Section III, "Rules for Construction of Nuclear Power Plant Components,"

7. American Concrete Institute and American Society of Mechanical Engineers (Joint Committee), "Code for Concrete Reactor Vessels and Containments," ACI 359, (also designated as ASME Boiler and Pressure Vessel Code, Section III, "Rules for Construction of Nuclear Power Plant Components," Division 2)

8. American Concrete Institute, "Building Code Requirements for Reinforced Concrete," ACI 318

9. American Concrete Institute, "Code Requirements for Nuclear Safety Related Concrete Structures ACI 349

10. U.S. Nuclear Regulatory Commission, "Design of an Independent Spent Fuel Storage Installation (Dry Storage)," Regulatory Guide 3.60

11. ANSI N14.6. "American National Standards for Radioactive Material Lifting Devices for Sipping Containers Weighing 10,000 lbs (4500 kg) or More"

12. American Institute of Steel Construction, "Specification for Structural Steel Buildings — Allowable Stress Design and Plastic Design," published in the AISC "Manual of Steel Construction"

13. American Institute of Steel Construction, "Load and Resistance Factor Design Specification for Structural Steel Buildings"

14. American Welding Society, "Structural Welding Code Steel," AWS D1.1. [This should be cited as applicable for structures that are subject to certification and are not within the specific scope of the ASME B&PV Code for the confinement cask.]

15. American Society of Civil Engineers, "Minimum Design Loads for Buildings and Other Structures," ASCE 7

16. American Concrete Institute, "Code Requirements for Nuclear Safety Related Concrete Structures," ACI 349-85, Appendix B

17. International Conference of Building Officials (ICBO) "Uniform Building Code", (UBC)

18. American Institute of Steel Construction, "Code of Standard Practice for Steel Buildings and Bridges," published in the AISC "Manual of Steel Construction"

19. American Society of Mechanical Engineers, Boiler and Pressure Vessel Code, Section VIII, "Rules for the Construction of Pressure Vessels"

20. U.S. Nuclear Regulatory Commission, "Fracture Toughness Criteria of Base Material for Ferritic Steel Shipping Cask Containment Vessels with a Maximum Wall Thickness of 4 inches (0.1m)," Regulatory Guide 7.11, June 1991

21. U.S. Nuclear Regulatory Commission, "Fracture Toughness Criteria of Base Material for Ferritic Steel Shipping Cask Containment Vessels with a Wall Thickness Greater than 4 inches (0.1m)," Regulatory Guide 7.12, June 1991

22. W.R. Holman, *et al.*, "Recommendations for Protecting Against Failure by Brittle Fracture in Ferritic Steel Shipping Containers Up to 4 Inches Thick," NUREG/CR-1815, Lawrence Livermore National Laboratory, June 1981

23. U.S. Nuclear Regulatory Commission, "Design Criteria for the Structural Analysis of Shipping Cask Containment Vessels," Regulatory Guide 7.6, March 1978

24. American Welding Society, "Standard Symbols for Welding, Brazing, and Nondestructive Examination," AWS A2.4

25. U.S. Nuclear Regulatory Commission, "A study on Ductile and Brittle Failure Design Criteria for Ductile Cast Iron Spent-Fuel Shipping Containers" NUREG-3760, January, 1986

26. NRC bulletin 96-04 "Chemical, Galvanic, or other Reactions in Spent Fuel Storage and Transportation Casks", July 1996

27. *U.S. Code of Federal Regulations*, Part 50, "Domestic Licensing of Production and Utilization Facilities," Title 10, "Energy"

28. *U.S. Code of Federal Regulations*, Part 71, "Packaging and Transportation of Radioactive Materials," Title 10, "Energy"

29. S.F. Hoerner, *Fluid-Dynamics Drag*, Hoerner Fluid Dynamics, P.O. Box 342 Brick Town NJ, 1965

30. U.S. Nuclear Regulatory Commission,"Design-Basis Tornado for Nuclear Power Plants," Regulatory Guide 1.76, April 1974

31. U.S. Nuclear Regulatory Commission, "Standard Review Plan for Power Reactors," NUREG-0800

32. W.B. Cottrell and A.W. Savolainen, Oak Ridge National Laboratory, "U. S. Reactor Containment Technology," ORNL-NSIC-5, Vol. 1, Chapter 6

33. R.P. Kennedy, *Review of Procedures for the Analysis and Design of Concrete Structures to Resist Missile Impact Effects*, Holmes and Narver Inc., Sept. 1975

34. R.J. Roark, *Formulas for Stress and Strain*, McGraw Hill, 1965

35. J.R. Stokley and D.H. Williamson, "Structural Integrity of Spent Nuclear Fuel Storage Casks Subjected to Drop," *Nuclear Technology*, Volume 114, Number 1, April, 1996.

36. U.S. Nuclear Regulatory Commission, "Control of Heavy Loads at Power Plants NUREG-0612, July 1980

37. A.S. Lee, S.E. Bumpas, Buckling Analysis if Spent Fuel Basket, NUREG/CR-6322, Lawrence Livermore National Laboratory, May, 1995

38. G.C. Mok, L.E. Fischer, Lawrence Livermore National Laboratory, S.T. Hsu, Kaiser Engineering "Stress Analysis of Closure Bolts for Shipping Casks," NUREG/CR-6007, January 1993

39. American Society of Mechanical Engineers, Boiler and Pressure Vessel Code, Section IX, "Welding and Brazing Qualifications"

40. U.S. Nuclear Regulatory Commission "Design Control for ISFSI Components", Inspection Procedure 60851

41. U.S. Nuclear Regulatory Commission, "Design of an Independent Spent Fuel Storage Installation" Regulatory Guide 3.60, March 1987

42. National Fire Protection Association, "Electric, Life Safety, and Lightning Protection Codes"

43. U.S. Nuclear Regulatory Commission, "Applicability of Existing Regulatory Guides to the Design and Operation of an Independent Spent Fuel Storage Installation," Regulatory Guide 3.53, July 1982

44. American Society of Civil Engineers, "Seismic Analysis of Safety-Related Nuclear Structures," ASCE 4

45. American Society of Mechanical Engineers, "Quality Assurance Program Requirements for Nuclear Facilities" ASME NQA-2

46. American National Standards Institute, American Nuclear Society, "Nuclear Facilities — Steel Safety-Related Structures for Design Fabrication and Erection," ANSI/AND N690

47. Nuclear Regulatory Commission, Regulatory Guide 1.26, U.S., "Quality Group Classification and Standard for Water-,Steam-, and Radioactive-Waste-Container Components of Nuclear Power Plants"

48. U.S. Nuclear Regulatory Commission, "Standard Review Plan For Nuclear Power Plants" NUREG-0800 Section 3.2.2

49. American National Standards Institute, American Society of Mechanical Engineers, "Power Piping," ANSI/ASME B31.1

50. American National Standards Institute, "ANSI B16.34 "Valves", B31.1"Power Piping" 1973

51. American Water Works Association, "Standard for Steel Tanks — Stand pipes, Reservoirs, and Elevated Tanks for Water Storage," AWWA D100

52. American National Standards Institute, American Society of Mechanical Engineers, "Specification for Welded Aluminum-Alloy Field-Erected Storage Tanks," ANSI/ASME B96.1

53. American Petroleum Institute, "Recommended Rules for Design and Construction of Large, Welded, Low-Pressure Storage Tanks," API 620

54. American Concrete Institute, "Building Code Requirements for Masonry Structures," ACI 530

4.0 THERMAL EVALUATION

I. Review Objective

The thermal review ensures that the cask and fuel material temperatures of the dry cask storage system (DCSS) will remain within the allowable values or criteria for normal, off-normal, and accident conditions. This objective includes confirmation that the temperatures of the fuel cladding (fission product barrier) will be maintained throughout the storage period to protect the cladding against degradation that could lead to gross rupture. This portion of the DCSS review also confirms that the thermal design of the cask has been evaluated using acceptable analytical and/or testing methods.

II. Areas of Review

This portion of the DCSS review evaluates the design and analysis of cask thermal performance for normal, off-normal, and accident conditions. Consequently, this chapter of the DCSS Standard Review Plan (SRP) provides guidance for use in reviewing thermal design criteria, design features, model specifications, and material properties. In addition, this chapter provides guidance for evaluating thermal analysis methods, including computer programs, temperature and pressure calculations, confirmatory testing, and independent evaluations done by staff.

As described in Section V, "Review Procedures," a comprehensive thermal evaluation *may* encompass the following areas of review:

1. spent fuel cladding
2. cask system thermal design
   a. design criteria
   b. design features
3. thermal load specification/ambient temperature
4. model specification
   a. configuration
   b. material properties
   c. boundary conditions
5. thermal analysis
   a. computer programs
   b. temperature calculations
   c. pressure analysis
   d. confirmatory analysis
6. supplemental information

III. Regulatory Requirements

10 CFR Part 72 requires an analysis and evaluation of DCSS thermal design and performance to demonstrate that the cask will permit safe storage of the spent fuel for a minimum of 20 years. The spent fuel cladding must be protected against degradation that may lead to gross ruptures. Thermal structures, systems, and components important to safety must be described in sufficient detail to permit evaluation of their effectiveness. Applicable thermal requirements are identified, in part, in 10 CFR 72.24(c)(3), 72.24(d), 72.122(h)(1), 72.122(l), 72.128(a)(4), 72.236(f), 72.236(g), and 72.236(h).

IV. Acceptance Criteria

In general, the DCSS thermal evaluation seeks to ensure that the given design fulfills the following acceptance criteria:

1. Fuel cladding (zircalloy) temperature at the beginning of dry cask storage should generally be below the anticipated damage-threshold temperatures for normal conditions and a minimum of 20 years of cask storage (Refs. 12 and 13).

2. Fuel cladding (zircalloy) temperature should generally be maintained below 570 °C (1058 °F) for short-term accident conditions, short-term off-normal conditions, and fuel transfer operations (e.g., vacuum drying of the cask or dry transfer). (PNL-4835[1])

3. The maximum internal pressure of the cask should remain within its design pressures for normal, off-normal, and accident conditions assuming rupture of 1 percent, 10 percent, and 100 percent of the fuel rods, respectively. Assumptions for pressure calculations include release of 100 percent of the fill gas and 30 percent of the significant radioactive gases in the fuel rods.

4. Cask and fuel materials should be maintained within their minimum and maximum temperature criteria for normal, off-normal, and accident conditions in order to enable components to perform their intended safety functions.

5. For each fuel type proposed for storage, the DCSS should ensure a very low probability (e.g., 0.5 percent per fuel rod) of cladding breach during long-term storage.

6. Fuel cladding damage resulting from creep cavitation should be limited to 15 percent of the original cladding cross-sectional area during dry storage. (UCID-21118)

7. The cask system should be passively cooled. [10 CFR 72.236(f)]

8. The thermal performance of the cask should be within the allowable design criteria specified in SAR Section 2 (e.g., materials, decay heat specifications) and SAR Section 3 (e.g., thermal stress analysis) for normal, off-normal, and accident conditions.

V. Review Procedures

One of the most important results of the DCSS thermal evaluation is confirmation that the fuel cladding temperature will remain sufficiently low to prevent unacceptable degradation during storage. The SAR should identify the allowable temperature levels for stored materials for long-term storage, as well as short-term storage and off-normal conditions.

Design features and criteria, initially presented in SAR Sections 1 and 2, should be reviewed for additional insight. Reviewers should examine heat loads from both contents and external sources, and should assess models used by the applicant for thermal analysis. Temperatures and pressures calculated in the SAR should be confirmed to evaluate compliance with design criteria and regulatory requirements.

Reviewers should also evaluate temperature distributions and criteria used to determine thermal stresses for all cask system components exposed to heat generated by the fuel. These components include the cask, transfer equipment, and any shielding components.

Temperatures and related distributions and pressures developed within the confinement cask should also be evaluated. (These values are further referenced and evaluated in Chapter 3, "Structural Evaluation," and Chapter 7, "Confinement Evaluation," of this SRP.) However, this thermal review does not encompass the evaluation of the computed stresses or loads on parts of the structure caused by either significant temperature gradients or the interaction of those materials at different temperatures each with possibly different coefficients of thermal expansion. These computations are reviewed as part of the structural evaluation discussed in Chapter 3 of this SRP.

Thermal performance of the cask under accident conditions, as evaluated in this portion of the DCSS review, is also addressed as appropriate in the overall accident analyses presented in Chapter 11. In conducting a comprehensive thermal evaluation, reviewers should perform the established review procedures, as applicable, for each of the following areas of review:

1. Spent Fuel Cladding

The cask system must be designed to prevent degradation of fuel cladding that results in a type of cladding breach, such as axial-splits of ductile fracture, where irradiated UO_2 particles may be released. In addition, the fuel cladding should not degrade to the point where more than one percent of the fuel rods suffer pinhole or hairline crack type failure under normal storage conditions. The one percent failure assumption is for safety analysis purposes only, and relates to assumptions for thermal analysis, containment performance, and cask unloading operations. However, the cask design should ensure a very low probability of rod failure during normal storage. "Failed fuel" (failure occurring during dry storage) us defined as fuel with a breach of cladding from but not limited to, pinhole failure, hairline cracks, axial-splits, or ductile fracture.

The staff should verify that cladding temperatures for each fuel type proposed for storage will be below the expected damage thresholds for normal conditions of storage. Zircalloy fuel cladding temperature limits at the beginning of dry storage are typically below 380 °C for 5-year cooled fuel and 340 °C for 10-year cooled fuel for normal conditions and a minimum of 20 years of cask storage (PNL-4835, PNL-6189, PNL-6364[2], and LLNL UCID-21181). Currently, the staff accepts only the diffusion-controlled cavity growth (DCCG) method for establishing the temperature limits, with DCCG damage not to exceed 15 percent of the original cladding cross-sectional area during dry storage (UCID-21181).

However, it should be noted that fuel cladding temperature limits are a complex function of power history (including transients), cladding thickness, pre-pressurization of fuel rods during fabrication, burnup, fission gas, and hoop stress. Substantial variation in end-of-life internal rod pressures and fuel design characteristics may warrant temperature limits lower than those noted above for certain fuel types. Therefore, fuel cladding limits for each fuel type should be defined in the SAR with thermal restrictions in the DCSS technical specifications.

Evaluate the method used to determine the temperature limits and associated cladding hoop stresses. Hoop stress calculations should be established on the basis of fuel properties that are representative of the spent fuel to be stored (e.g., cladding dimensions, internal rod pressures). High-burnup of fuel (greater than 40,000 MWD/MTU) causes effects, such as wall thinning from increased oxidation and increased internal rod pressure from fission gas buildup, and changes in fuel dimensions that must be evaluated. The SAR should use conservative values for surface oxidation thickness. Oxidation may not be of a uniform thickness along the qxial length of the fuel rods and average values may under predict wall thinning. Temperature limits will be more restrictive with increased fuel cooling time (and/or increased burnup), largely as a result of creep cavitation.

For short term-accident conditions, the staff accepts zircalloy fuel cladding temperature generally maintained below 570 °C (1058 °F). The short-term accident temperature of 570 °C (1058 °F) for zircalloy-clad fuel is currently accepted as a suitable criterion for fuel transfer operations. However, this temperature limit may be lowered for fuel with hoop stresses exceeding the rods that were high-temperature tested (see Table 5 of PNL-4835). This is especially true for fuel with burnup greater than the tested rods (e.g., greater than ~28,000 MWD/MTU), as a result of increased internal rod pressure from fission gas buildup and release into the gap and/or different helium loadings. The applicant should verify that these cladding temperature limits are appropriate for all fuel types proposed for storage, and that the fuel cladding temperatures will remain below the limit for facility operations (e.g., fuel transfer) and the worst-case credible accident.

For cask unloading operations, cladding integrity should be maintained during reflooding, so as not to interfere with fuel handling and retrieval. The SAR should include a quench analysis supporting specified minimum quench fluid temperature and maximum fluid flow rate during re-flood. This analysis should also be referenced in Section 11 of the SAR as having been considered in the developing the model unloading procedures, and be included as appropriate in the Technical Specifications for the system use. The NRC accepts the fact that the total stress on the cladding must be maintained below the material's minimum yield stress. The total stress includes the thermal stress combined with the cladding hoop stress from internal rod pressure and rod gas plenum temperature. The analysis should account for high burnup effects on the fuel (e.g., waterside corrosion, high internal rod pressure) and minimum manufacturing wall thickness.

Verify that the applicant includes technical specifications which ensure that the maximum allowed initial cladding temperatures will not be exceeded during normal operations. Technical Specifications should also identify the maximum time permitted for fuel to be submerged in a cask which has been removed from the pool (i.e. when the cask has not been evacuated and sealed). A series of curves for cladding temperature versus time for differing decay heat payloads is acceptable.

2. Cask System Thermal Design

a. Design Criteria

Review the principal design criteria, as well as the structure, system, and component specifications presented in the SAR Section 2 and any additional detail provided in Section 4.

b. Design Features

Review the description of the significant thermal design features and operating characteristics of pertinent DCSS subsystems. Design features typically include the cask body, thermal fins, shielding materials, fuel baskets, impact limiters if installed, containment seals, drain and vent ports, and pressure relief devices, among others. Verify that the thermal design features will adequately perform their intended safety functions during normal, off-normal, and accident conditions.

All thermal design features should be passive. Review the cask system component specifications for inclusion of material composition and thermal properties, operating pressures and criteria for any relief devices or rupture disks. Review the general description of the cask presented in SAR Section 1, as supplemented by the additional information provided in SAR Section 4. In addition to the material compositions, dimensions of the cask components, and spacing of fuel assemblies in the basket, the thermal design may include external air passages. Rupture disks may be used on cask components, such as shielding, to ensure that elevated temperatures do not cause thermal expansion or phase changes resulting in structural damage to the cask shell. All drawings, figures, and tables should be sufficiently detailed to support in-depth staff evaluation.

Any instrumentation used to monitor cask thermal performance should also be described in sufficient detail to support in-depth staff evaluation. The monitoring instrumentation components should have a safety classification commensurate with their function, and the safety classification (presented in SAR Section 2) should be justified. Applicable operating controls and criteria, such as temperature criteria and surveillance requirements, should be clearly indicated in SAR Section 12, discussed in the SER, and included in the license or certificate of compliance, as appropriate.

3. Thermal Load Specification/Ambient Temperature

Examine the specification for the design-basis fuel decay heat presented in SAR Section 2. Ensure that this decay heat is consistent with the specified burnup and cooling times, if included. Decay heat is generally calculated using the same computer codes as those used to determine radiation source terms. Coordinate the review of fuel source terms for consistency with the shielding review, as appropriate. Alternatively, the decay heat from the design-basis fuel may also be derived from Regulatory Guide (RG) 3.54.[3] Except for neutrino energy, all decay heat should be considered to be deposited in the fuel.

If control components or other assembly hardware (e.g., shrouds) are included with the fuel assemblies, their heat loads should be specified and justified.

In general, the NRC staff accepts insolance presented in 10 CFR Part 71[4] for 10 CFR Part 72 applications. Because of the large thermal inertia of a storage cask, the values listed in 10 CFR Part 71.71 may be treated as the average insolance, calculated by averaging over a 24-hour day the reported 10 CFR 71 values for insolance over a 12-hour solar day, in a steady-state calculation. If a less conservative approach is presented, the SAR shall thoroughly describe and justify its use.

Review the ambient temperatures used to calculate temperature distributions and maxima for normal and off-normal conditions, as well as "design-basis" natural phenomena. The SAR should clearly state the assumed temperatures and temperature variations over time. These assumptions establish criteria for comparison with recorded data and projections for potential installations of the cask system at specific sites.

When calculating maximum thermal gradients and temperature differences within individual components or between locations, temperature changes over time may need to be determined. These changes should consider the types of material, the thermal properties, including thermal conductivity; heat capacity; and density of specific components. Statements about assumed bounding temperature ranges, ambient temperature conditions, and variations of external heat sources over time should be defined so that they may be easily compared with available site or regional data.

For those cask system components for which material properties and performance vary with temperature, review the assumptions used in determining temperature maxima, minima, gradients, and differences for the cask system. Also review the assumptions used to determine fuel cladding temperatures. The assumed temperature changes over time should result in the bounding conditions for the structural analysis. The calculated temperatures in the various cask system components should be compared to the limiting temperature criteria for the appropriate materials. Ferritic materials are subject to failure by brittle fracture at low temperatures. Review the assumed low temperatures for cask system handling operations for consistency with material properties. Ambient temperature restrictions may be appropriate for cask handling operations. Any limiting conditions regarding ambient temperatures should be addressed in SAR Section 12 as well as SER Section 12, and should be included as a limiting condition of operation (e.g. tech. spec.)in the license or certificate of compliance, as appropriate.

During wet fuel transfer operations, the liquid in the cask should not be permitted to boil. This practice avoids uncontrolled pressures on the cask and the connected dewatering, purging, and recharging system(s), unacceptable discharge of liquids which may be providing radiation shielding, and a potentially unacceptable reduction in the safety margin (K_{eff}) that prevents inadvertent criticality. The reviewer should ensure that to prevent any of the above conditions, an adequate subcooling margin is identified in both the SAR and corresponding operating procedures to prevent boiling. This margin may be cask-specific, depending on the design of the fuel basket and key assumptions used in the criticality analysis. Review the heatup and time-to-boil calculations and assess whether any technical specification or limiting conditions for operation are needed. Heatup calculations should be established on the basis of the spent fuel pool's technical specification maximum temperature limit (typically 115 °F).

If the fuel cladding temperature calculation is based on heatup over a limited time period for cask drying operations, verify that limiting conditions for the operations have been imposed in the technical specifications. Such limiting conditions should ensure that the temperature will remain acceptable during the operations, and that normal cooling will begin before the temperature criterion is exceeded.

For unloading operations, evaluate temperature and pressure calculations supporting procedural steps presented in SAR Section 8 for cask cooldown and reflooding of the cask internals. To ensure that the cask does not overpressurize and that the fuel assemblies are not subjected to excess thermal stresses, the applicant's analysis should specify and justify the appropriate temperature and flow rate of the quench fluid, assuming maximum fuel cladding temperatures in the unloading configuration. The NRC accepts that the total stress on the cladding is maintained below the material's minimum yield stress (see the spent fuel cladding review procedures in this section). Other assembly components should also be examined in a similar manner. Engineering judgement combined with relevant industry operational experience with unloading spent fuel from transportation and storage casks may support the basis for limits on quench fluid temperature and flow rate. This review should be coordinated with the structural review (SRP Chapter 3) and procedure review (SRP Chapter 8).

Analysis for accident-level ("design-basis") temperatures should not be considered to envelop the analysis of normal or off-normal temperatures. Normal and off-normal temperature demands for structural capacity have different acceptance criteria. Therefore, all three conditions should be analyzed. In addition, the duration over which accident temperature conditions may exist should be evaluated. Because material properties may be functions of temperature and time, long-term temperature elevations can cause gradual degradation of material properties.

4. Model Specification

a. Configuration

Verify that the model used in the thermal evaluation is clearly described. Separate models may be used for the evaluation of normal and accident conditions. Coordinate with the structural review (SRP Chapter 3) to evaluate any damage that may result from accidents or natural phenomena events. All models should be shown to be conservative.

Examine the sketches or figures of the model used for thermal calculations. Verify that the dimensions and materials of the model are consistent with those in the drawings of the actual cask, as presented in SAR Section 1. If possible, examine computer inputs to verify consistency with the model sketches and engineering drawings. Differences between the actual cask configuration and the model should be identified, and the model should be shown to be conservative.

Pay particular attention to gaps between cask components. Tolerances should be considered so that the thermal resistance of each gap is treated conservatively. Gases (e.g., air, helium) assumed to be present in the gap shall be described and justified. If a specific gas other than air in the cask cavity or gaps between cask components is relied upon for heat removal, the applicant should show that the gas is retained *and* that the gas is not diluted by other gases having lower thermal conductivities during the entire storage period. For cask components that are important to heat removal, manufacturing techniques for joining components, surface roughness, contact pressures, and gap conductance values should be adequately described and justified.

Review the decay heat load axial distribution. Ensure that decay heat generated in the spent fuel is limited to the active fuel region of the assemblies. The model should specifically account for the peaking in the central region. Heat from control components, if applicable, should also be distributed appropriately. In addition, the positions of heat sources relative to other cask components should be identified.

Examine the heat transfer processes used in the analysis. Conduction and radiation are typically defined as the primary heat transfer mechanisms within the cask itself. Convection by natural circulation should be limited to that between the external surface of the cask and the ambient environment. The staff has not previously approved specific thermal models for natural circulation internal to the cask because of the difficulty in modeling and the lack of test data. Applicants seeking NRC approval of specific internal convection models should propose, in the SAR, a comprehensive test program to demonstrate the adequacy of the cask design and validation of the convection models. Actual spent fuel properties and uncertainties (e.g., friction factors, crud and oxide buildup, eccentricities, non-uniform axial and radial decay heat profiles) should also be addressed. Applicants using an effective thermal conductivity for the cover gas (e.g., helium) in lieu of a specific convection model should also justify values used in the analysis.

Use of effective thermal conductivity coefficients for regions within the confinement cask other than the fuel (e.g., gaps) may overestimate heat transfer. If effective thermal conductivity is used in this manner, verify that the same values have been determined from test data that are representative of similar geometry, materials, temperatures, and heat fluxes used in current application. Pay particular attention to the effective thermal conductivity of neutron shield regions, such as those embedded with thermal fins. Voids or gaps typically exist, as a result of either tolerances or shrinkage, and shall be considered in calculating effective thermal conductivity. Also, pay particular attention to the values assumed for surface emissivities and view factors, as well as the manner used to account for radiation heat transfer in determining the effective thermal conductivities.

Coordinate the thermal review with the structural review (SRP Chapter 3) to ensure that, for components external to the confinement cask, the applicant analyzed situations that may produce the worst-case cask loads. As an illustration, for cask systems that may have multiple shielded casks and/or ones that provide cooling air passages by a single, integral structure, the greatest gradients and loadings caused by thermal expansion may occur with casks in alternative storage or in temporary handling positions, whereas the highest material temperatures probably occur with casks in any position.

Review how the SAR treats heat transfer through the fuel assemblies and, if applicable, the manner in which effective conductivity is determined for each fuel assembly. The fuel may be modeled as a homogenous region using an effective thermal conductivity. The basket wall temperature of the hottest assembly can then be used to determine the peak rod temperature of the hottest assembly using the Wooten-Epstein correlation. Guidance on effective thermal conductivity of the fuel is presented below in the discussion concerning material properties.

Verify that the SAR addresses the thermal interaction among casks in an array by using a view factor less than unity. Generally, this will result in an operating control and limit in SAR Section 12 that imposes a minimum spacing between storage casks.

b. Material Properties

Verify that the material compositions and thermal properties are provided for all components used in the calculational model. Verify that the thermal properties used in the safety analysis are appropriate, and that potential degradation of materials over their service life has been evaluated. The source of the thermal property data should be traced to an authoritative reference (generally not a textbook). The NRC has accepted the American Society of Mechanical Engineers (ASME) Boiler and Pressure Vessel (B&PV) Code, Division 1, Section II, "Material Specifications," and Section III[5] appendices as a primary source for material properties. Pay particular attention to non-standard materials (e.g., neutron shielding and seals). Temperature and anisotropic dependencies of thermal properties should be considered. In addition, if regional thermal properties are determined from a combination of individual materials, the manner in which these effective properties are calculated should also be described.

If the transverse effective thermal conductivity of the fuel is greater than 0.5 BTU/hr-ft-°F (~0.86 W/m-K) under the conditions described in Effective Thermal Conductivity and Edge Configuration Model for Spent Fuel Assembly[6], the method in which it was determined shall be thoroughly described and supported. If the thermal model is axisymmetric or three-dimensional, the longitudinal thermal conductivity should generally be limited to the conductivity of the cladding (weighted by its fractional area) within the fuel assembly. Gaps between fuel pellets and cracks in the pellets themselves can result in a considerable uncertainty regarding the contribution of the fuel to longitudinal heat transfer. High burnup effects should also be considered in determining the fuel region effective thermal conductivity.

The SAR should also indicate both maximum and minimum temperature criteria for cask materials and components, and the justification and references for these criteria shall be adequately described. Criteria for concrete temperatures are established by American Concrete Institute (ACI) Standard 359[7] for structures, systems, and components within the scope of that code, as well as Appendix A to ACI 349[8].

c. Boundary Conditions

The applicant should identify boundary conditions for normal, off-normal, and fire accident conditions. The required boundary conditions include the decay heat rate from each fuel assembly and the external conditions on the cask surface. The peak power factor for a fuel assembly should be specified and the peak linear power of a fuel assembly should be stated for a given active fuel length. The peak decay heat flux on a basket compartment surface should also be given.

The boundary conditions on the cask surface depend on the environment surrounding the cask. Consequently, the temperature of the environment should be specified for normal and off-normal conditions, as should the incident and absorbed insolance. The mechanisms and models for dissipating the absorbed insolance and decay heat from the surface of the cask to the environment should also be identified and described. The mechanisms for transferring heat from the cask surface during normal and off-normal conditions usually consist of natural (free) convection and thermal radiation. A heat balance on the surface of the cask should be given and the results presented.

The initial temperature distribution of the storage cask system before a fire accident should be established on the basis of the hottest temperature distribution during normal or off-normal storage conditions. The duration and flame temperature of the fire should be specified, as should the flame velocity and emissivity. The flame and cask surface emissivities specified in 10 CFR 71.73(a)(3) (April 1996) for a hypothetical accident test of transportation packages are satisfactory for use with regard to a fire accident involving a storage cask.

The applicant should identify and describe the mechanisms and models for coupling the fire energy to the cask surface. These mechanisms include forced convection in relation to the flame velocity (5-15 m/s) as well as thermal radiation. In addition, the applicant should justify convection coefficients during the fire. Natural convection coefficients are not appropriate, as such coefficients imply downward gas flow adjacent to relatively cool cask walls. In general, buoyant upward flow will dominate.

Following the fire, the cask is subject to insolation and content decay heat while being cooled by natural convection and thermal radiation to the environment. The applicant should identify the post-fire conditions of the cask, including any changes in surface conditions and/or geometry that may affect radiation and convection heat losses. In addition, the applicant should identify and describe the models for the post-fire processes.

5. Thermal Analysis

a. Computer Programs

Determine which computer codes were used in the thermal evaluation. The applicant should use well-verified and validated computer code used to perform the thermal evaluation. The two codes most frequently encountered in SARs are ANSYS[9] and HEATING[10]. Both are capable of general 3-D steady state and transient calculations. Assess that the number of dimensions and temporal treatment are appropriate for the calculations being performed.

At least two codes, SCANS[11] and CASKS[12], have been developed to perform simple, approximate confirmatory analyses. These codes are not acceptable for use in an SAR for thermal design and analysis. In addition, since these codes address temperatures of only the cask body, they cannot be used as the sole confirmatory tool for the thermal review.

The SAR documentation should include input and output file listings for the thermal evaluations. Reviewers should be familiar with the codes used in the SAR documentation. If the applicant proposes to use codes not previously accepted by NRC, development of reviewer familiarity with those codes is considered to be a necessary part of the review process. The applicant should also describe, in the SAR, the code and justification for use in the thermal evaluation. Verify that the information from the thermal model is properly input into the code. Verify that the output has been properly interpreted and applied in the thermal and structural analyses. The scope of confirmatory calculations is partly dependent on the quality of the output data and its use.

b. Temperature Calculations

The SAR should include a table that lists the maximum and minimum temperatures of all components important to safety under normal, off-normal, and accident conditions. This table should specify the operating temperature range for each component. Verify that temperatures have been calculated for key components and that they do not exceed the allowable range for each. Justification shall be provided in the SAR for any material important to safety that exceeds acceptable temperature ranges. If compliance with minimum temperature criteria relies on a specific minimum heat load from the fuel, such heat load shall be quantified and included as an operating control and a technical specification criterion in SAR Section 12.

Pay particular attention to the maximum temperature of the cladding. Currently, the staff accepts temperature criteria established on the basis of the DCCG[13]. Comparable criteria are also defined in PNL-6189[14], even though the maximum temperatures are established on the basis of a different failure mechanism (creep). Experience with previous SARs has shown that most of the review effort is generally devoted to confirming that these temperature criteria are satisfied.

Some storage systems rely upon natural circulation of air through internal passages to remove heat from the stored confinement cask. For storage systems with internal air flow passages, blockage of inlet and/or outlet flow is an accident situation that should be evaluated. Total blockage of all inlets and outlets may result in fuel heatup, which has been assumed to approach adiabatic conditions. To ensure that blockages do not go undetected for significant periods, the NRC has required objective evidence that inlet and outlet flows are not obstructed. Consequently, for these type of storage systems, the NRC has accepted periodic visual inspection of the vents coupled with temperature measurements to verify proper thermal performance and detect flow blockages. The inspections should take place within an interval that will allow sufficient time for corrective actions to be taken before the accident temperature is reached. The inspection interval should be more frequent than the time interval required for the fuel to heatup to the established accident temperature criteria, assuming a total blockage of all inlets and outlets.

Review of the heatup calculations should especially address any assumptions regarding limiting components and quasi-steady state responses. The initial ambient temperature for the heatup calculations should bound the maximum "normal condition" temperature. The resulting heatup time history should be included in the SAR documentation, and should support the proposed inspection and monitoring intervals. The information is also useful in developing contingency operation procedures, since it indicates the available time in which to take corrective actions before the fuel accident temperature criteria may be exceeded.

The most extreme thermal conditions may result from credible ambient temperatures, temperature-time histories, an adjacent fire, or any off-normal or design-basis event (DBE) resulting in blockage of ventilation passages. The worst-case structural loads may occur at temperatures lower than those of design-basis accidents or natural phenomena, since load combination expressions effectively require greater safety factors for normal and off-normal analyses than for design-basis (accident) events and conditions. For storage systems without internal air flow passages, the worst-case accident thermal conditions typically have been fire.

Burning of fuel and other combustibles associated with vehicles involved in transfer operations should, at a minimum, be presumed to be a DBE with the cask in the most exposed situation during transfer or loading into storage. Fire parameters included in 10 CFR 71.73 have been accepted for characterizing the heat transfer during the in-storage fire. However, a bounding analysis that limits the fuel source thus limits the length of the fire (e.g., by limiting the source to the fuel in the transporter) has also been accepted. If the SAR does not address fire, or if the site-specific fire parameters exceed those of the SAR, the site-specific application will need to include analysis of the worst-case credible fire.

Some structures, systems, and components may experience the most severe conditions if exposure to high temperatures is followed by dousing (as by rain or fire water). A small amount of exterior concrete spalling may result from a fire, the application of fire suppression water, rain on heated surfaces or other high-temperature condition. The damage from these events is readily detectable, and appropriate recovery or corrective measures may be presumed. Therefore, the loss of such a small amount of shielding material is not expected to cause a storage system to exceed the regulatory requirements in 10 CFR 72.106 and, therefore, need not be estimated or evaluated in the SAR. The NRC accepts that concrete temperatures may exceed the temperature criteria of ACI 349 for accidents if the temperatures result from a fire. In that case, corrective action may be required for continued safe storage.

The methods that are acceptable for analyzing and reviewing the consequences of a fire depend upon the duration of the fire and the margin between the predicted temperatures and the actual thermal limits of the components. For a fire of very short duration (i.e., less than 10 percent of the thermal time constant of the cask body), the NRC finds it acceptable to calculate the fuel temperature increase by assuming that the cask inner wall is adiabatic. The fuel temperature increase should then be determined by dividing the decay energy released during the fire by the thermal capacity of the basket-fuel assembly combination. For a fire of somewhat longer duration, it is acceptable to evaluate the cask body with no credit for the thermal capacity of the fuel assemblies and basket. This model assumes that the fuel temperature increase relates to the decay heat conducted or radiated from the fuel assemblies to the maximum temperature of the cask inner wall. A fire of sufficient duration, or one in which material temperatures are close to the criteria of their acceptable operational range, will require a detailed model of the cask and its contents. Cask system components (e.g., the neutron shield) may be assumed to be intact at the start of the fire, unless the fire is a secondary effect resulting from another credible event that may have physically affected the cask system (e.g., an aircraft impact).

Some storage systems may use a transfer cask to move the loaded confinement cask to the independent spent fuel storage installation (ISFSI) storage site. When the confinement cask is within the transfer cask, cooling is typically less than for normal operation. Fuel cladding temperatures would therefore be expected to be higher than for normal storage conditions. This is generally acceptable as long as the short-term accident temperature of 570 °C (1058 °F) is not exceeded.

Examine the temperature distribution calculations for the fuel container inside the transfer cask. Verify that heat transfer through gap regions has been treated in a conservative manner, and that material properties and dimensions of the transfer cask are consistent with the design data defined in the SAR documentation. The initial ambient temperature should be the maximum "normal condition" temperature. Cask preparation for storage or unloading operations may include situations in which the confinement cask is evacuated while it is in the transfer cask. For such conditions, determine that the maximum fuel cladding temperature has been calculated in an acceptable manner. The short-term accident temperature of 570 °C (1058 °F) for zircalloy fuel cladding has been accepted as a suitable criterion for fuel transfer operations, with restrictions (see the earlier discussion of spent fuel cladding review procedures). If the calculation is based on heatup over a limited time, verify that limiting conditions for the operations have been imposed to ensure that the temperature will remain acceptable during the process, and that normal cooling will begin before the temperature criterion is exceeded.

If a cask tipover is a credible accident, verify that the applicant has evaluated the effect on cask and fuel temperatures in the new configuration. An analysis may be warranted when a significant portion of heat removal capability is attributed to internal convection if a change in orientation of that cask may have a significant effect.

Using the accident condition temperatures, verify that the applicant has correctly determined the post-accident pressure of the gas in the cask cavity. The pressure should be determined on the basis of the assumption that 100 percent of the fuel rods have failed. The resulting load on the cask confinement boundary should be used in the structural analysis with the appropriate load combinations for accident conditions.

c. Pressure Analysis

Verify that the containment pressure of the cask is within its design limits for normal and accident conditions. Pressure calculations should be performed using the ideal gas law (i.e., $PV = nRT$) and summing the partial pressures of each of the gas constituents in the cask cavity. The SAR should identify the method and all assumptions used in the pressure analysis. The SAR should also identify and justify the method used to determine the fission gas inventory. Reviewers should also assess the applicant's calculation packages.

In addition, it is necessary to consider the temperature distribution of all components within the cask cavity and the cavity walls in calculating the gas pressure in the cavity. For fire analysis, the SAR should identify the maximum gas temperature reached during the post-fire transient phase, explain the method used to determine the average gas temperature, and specify the time in the transient at which the peak gas temperature is attained.

This pressure also depends on the free volume in the cask cavity, the amount (moles) of cover gas (helium) in the cavity, and the amount of gases released from ruptured fuel pins. The free volume calculation should be reviewed to determine if all components internal to the cask cavity (e.g., fuel assemblies, basket, structural supports, spacer disks, reactor control components) have been properly considered. Free volume calculations should account for thermal expansion of the cask internal components and the fuel when subjected to accident temperatures.

The NRC accepts that normal-conditions occur with less than 1 percent of the fuel rods failed, off-normal-conditions occur with up to 10 percent of the fuel rods ruptured, and 100% of the fuel rods will have ruptured following a design-basis accident event. The NRC also accepts that a minimum of 100 percent of the fill gas and 30 percent of the significant radioactive gases (e.g., H^3, Kr, and Xe) within a ruptured fuel rod is available for release into the cask cavity. Verify that design criteria pressures stated in SAR Section 2 are consistent with the calculated maximum pressures. Verify that the pressure testing specified in SAR Section 9 is consistent with the calculated pressures.

d. Confirmatory Analysis

Reviewers should perform a confirmatory analysis of the thermal performance of the cask structures, systems, and components identified as important to safety. Review the SAR to ensure that the applicant made the correct assumptions and provided the correct input, and that the output is consistent with established physical (thermal) behavior. These results should specifically include steady-state temperature distributions; local heat balances; temperatures reached and temperature distributions within any reinforced concrete structures, systems, and components; and cask cavity pressures for the bounding ambient temperatures.

To provide the most reliable confirmation, confirmatory analysis should, to the degree possible, use a different thermal method than the code used by the applicant. Similar confirmation of transient temperatures (e.g., during a fire) should be performed, as applicable to the SAR analysis.

The minimum confirmatory review should include verifying that key design parameters have been appropriately determined and correctly expressed as input into the computer program(s) used for the thermal analysis. Key parameters include proper dimensions, material properties (including surface emissivities and view factors for radiation), and definition of heat sources. A heat balance at the outer surface of the cask should be performed to verify that the heat from the spent fuel and insolance, balance that removed by convection and radiation. Correlations for the heat transfer coefficient should then be assessed to confirm that they are appropriate for the existing storage conditions. The temperature of the cask's inner surface should be estimated by calculating the temperature distribution across the cask body with simple heat balance approximations. Finally, the difference between the cask's inner surface temperature and the maximum cladding temperature should be compared with that of similar casks and baskets reviewed in previous SARs.

If a more detailed confirmatory review is required, a portion of the cask or basket may be modeled and evaluated to ensure that the SAR results are realistic and conservative. An extensive confirmatory analysis is necessary if major errors are suspected, if the applicant's margin in a complex analysis is small, or if little conservatism exists in the SAR modeling approach. As an alternative, the applicant may be required to perform design-verification testing of an as-built cask system to confirm the thermal analyses presented in the SAR. Such testing may include verifying gap conductance values assumed in modeling thermal resistance. The test conditions, configuration, and type and location of instrumentation used, if any, should be sufficiently described in SAR Section 9.

The NRC accepts simplifying assumptions for the effects of reinforcing steel in determining the thermal performance and temperature distributions of reinforced concrete. Use of a homogeneous material, instead of modeling the concrete and reinforcing steel as separate elements, is acceptable if the substitute hypothetical material has appropriately adjusted the thermal properties, and the reinforcing steel is covered with concrete in accordance with the applicable structural code. More specific analysis may be required for thermal performance and/or temperature distributions of reinforced concrete designs with features that allow significant thermal transfer below the concrete surface (such as internal studs welded to an exposed steel plate).

6. Supplemental Information

Supplemental information may include copies of applicable references (if not generally available to reviewers), computer code descriptions, input and output files, and any other information that the applicant deems necessary. Likewise, reviewers should request any additional information needed to complete the evaluation process.

VI. Evaluation Findings

Review the 10 CFR Part 72 acceptance criteria and provide a summary statement for each. These statements should be similar to the following model:

Structures, systems, and components (SSCs) important to safety are described in sufficient detail in Sections _____ of the SAR to enable an evaluation of their thermal effectiveness. Cask SSCs important to safety remain within their operating temperature ranges.

The [cask designation] is designed with a heat-removal capability having verifiability and reliability consistent with its importance to safety. The cask is designed to provide adequate heat removal capacity without active cooling systems.

The spent fuel cladding is protected against degradation leading to gross ruptures by maintaining the cladding temperature for _____ -year cooled fuel below _____ °C in an [applicable gas] environment. Protection of the cladding against degradation is expected to allow ready retrieval of spent fuel for further processing or disposal.

The staff concludes that the thermal design of the [cask designation] is in compliance with 10 CFR Part 72, and that the applicable design and acceptance criteria have been satisfied. The evaluation of the thermal design provides reasonable assurance that the [cask designation] will allow safe storage of spent fuel for a licensed (certified) life of _____ years. This finding is reached on the basis of a review that considered the regulation itself, appropriate regulatory guides, applicable codes and standards, and accepted engineering practices.

VII. References

1.   A.B. Johnson and E.R. Gilbert, Pacific Nuclear Laboratories, "Technical Basis for Storage of Zircalloy-Clad Spent Fuel in Inert Gases," PNL-4835, September 1983.

2.   Cunningham, M.E. , et al "Control of Degradation of Spent LWR Fuel During Dry Storage in and Inert Atmosphere" PNL-6364, PNL September 1987

3.   U.S. Nuclear Regulatory Commission, "Spent Fuel Heat Generation in an Independent Spent Fuel Storage Installation," Regulatory Guide 3.54, September 1974.

4.   U.S. Code of Federal Regulations, "Packaging and Transportation of Radioactive Material," Part 71, Title 10, "Energy."

5.   American Society of Mechanical Engineers (ASME) Boiler and Pressure Vessel (B&PV) Code, Division 1, Section II, "Material Specification,"

6.   R.D. Manteufel and N.E. Todreas, "Effective Thermal Conductivity and Edge Configuration Model for Spent Fuel Assembly," *Nuclear Technology*, Vol. 105, pp. 421–440, March 1994.

7.   American Concrete Institute/American Society of Mechanical Engineers Joint Technical Committee, ACI 359 (ASME B&PV Code, Section III, Division IIA)

8.   American Concrete Institute, "Code Requirements for Nuclear Safety-Related Concrete Structures," ACI 349.

9.   Swanson Analysis Systems, Inc., "ANSYS Computer Code for Large-Scale General-Purpose Engineering Analysis," Houston, Texas

10.   K.W. Childs, Oak Ridge National Laboratory, "Heating 7.2 User's Manual," NUREG/CR-0200, Vol. 2, Rev. 4, April 1995.

11.   G.C. Mok, *et al.*, Lawrence Livermore National Laboratory, "SCANS [Shipping Cask Analysis System] — A Microcomputer-Based Analysis System for Shipping Cask Design Review," NUREG/CR-4554, 1989.

12.   T.F. Chen, *et al.*, Lawrence Livermore National Laboratory, "CASKS [Computer Analysis of Storage Casks]: A Microcomputer-Based Analysis System for Storage Cask Design Review," NUREG/CR-6242, February 1995.

13.   M.W. Schwartz and M.C. Witte, Lawrence Livermore National Laboratory, "Spent Fuel Cladding Integrity During Dry Storage," UCID-21181, September 1987.

14.   I.S. Levy, *et al.*, Pacific Northwest Laboratory, "Recommended Temperature Limits for Dry Storage of Spent Light-Water Zircalloy Clad Fuel Rods in Inert Gas," PNL-6189, May 1987.

5.0 SHIELDING EVALUATION

I. Review Objective

In this portion of the dry cask storage system (DCSS) review, the NRC evaluates the shielding features of the proposed cask system, as designed for an independent spent fuel storage installation (ISFSI). In conducting this review, the NRC staff seeks to ensure that the proposed shielding features provide adequate protection against direct radiation from the cask contents. The shielding features should limit the dose from direct radiation to the operating staff and members of the public, so that the dose remains within regulatory requirements during normal operating, off-normal, and design-basis accident (DBA) conditions.

II. Areas of Review

This chapter of the DCSS Standard Review Plan (SRP) provides guidance for use in evaluating the shielding features of the proposed cask system. As defined in Section V, "Review Procedures," a comprehensive shielding evaluation *may* encompass the following areas of review:

1. shielding design description
    a. design criteria
    b. design features
2. radiation source definition
    a. gamma source
    b. neutron source
3. shielding model specification
    a. configuration of shielding and source
    b. material properties
4. shielding analyses
    a. computer programs
    b. flux-to-dose-rate conversion
    c. dose rates
    d. independent calculations
5. supplementary information

As prescribed in 10 CFR Part 72[1], the regulatory requirements for doses at and beyond the controlled area boundary include both the direct dose and that from an estimated release of radionuclides to the atmosphere (based on the tested leak tightness of the confinement). Consequently, an overall assessment of the compliance of the proposed DCSS with these regulatory limits is contained in Chapter 10, "Radiation Protection," of this SRP.

III. Regulatory Requirements

10 CFR Part 72 requires that spent fuel radioactive waste storage and handling systems be designed with suitable shielding to provide adequate radiation protection under both normal and accident conditions. Consequently, the DCSS application must describe the shielding structures, systems, and components (SSCs) important to safety in sufficient detail to allow the NRC staff to thoroughly evaluate their effectiveness. It is the responsibility of the vendor, the facility owner, and the NRC staff to analyze such SSCs with the objective of assessing the impact of direct radiation doses on public health and safety.

In addition, SSCs important to safety must be designed to withstand the effects of both credible accidents and severe natural phenomena without impairing their capability to perform their safety functions. The applicable shielding requirements are identified, in part, in 10 CFR 72.24(c)(3), 72.24(d), 72.104(a), 72.106(b), 72.122(b), 72.122(c), 72.128(a)(2), and 72.236(d).

IV. Acceptance Criteria

The task of identifying dose rate limits for direct radiation from storage casks is complicated by three considerations. First, 10 CFR Part 72 states regulatory dose limits in terms of total absorbed doses rather than dose rates. Second, dose analyses must include potential sources of radiation other than direct radiation from spent fuel in the cask. Third, the regulatory requirements (listed below) for acceptable cask use at an ISFSI are site-specific and must be separately evaluated on a case-by-case basis. That is, these evaluations are performed as required for a site-specific license application or as required by 10 CFR 72.212 for a utility using a cask under the general license.

In general, the DCSS shielding evaluation seeks to ensure that the proposed design fulfills the following acceptance criteria:

1. The minimum distance from each spent fuel handling and storage facility to the controlled area boundary must be at least 100 meters. The "controlled area" is defined in 10 CFR 72.3 as the area immediately surrounding an ISFSI or monitored retrievable storage (MRS) facility, for which the licensee exercises authority regarding its use and within which ISFSI operations are performed.

2. The cask vendor must show that, during both normal operations and anticipated occurrences, the radiation shielding features of the proposed DCSS are sufficient to meet the radiation dose requirements in Sections 72.104(a). Specifically, the vendor must demonstrate this capability for a typical array of casks in the most bounding site configuration. For example, the most bounding configuration might be located at the minimum distance (100 meters) to the controlled area boundary, without any shielding from other structures or topography.

3. Dose rates from the cask must be consistent with a well-established "as low as reasonably achievable" (ALARA) program for activities in and around the storage site.

4. After a design-basis accident, an individual at the boundary or outside the controlled area shall not receive a dose greater than 5 rem to the whole body or any organ.

5. The proposed shielding features must ensure that the DCSS meets the regulatory requirements for occupational and radiation dose limits for individual members of the public, as prescribed in 10 CFR Part 20[2], Subparts C and D.

V. Review Procedures

1. Shielding Design Description

a. Design Criteria

Dose rates at the cask surface and in the vicinity of a loaded cask may vary during storage, transfer, and in-storage activities. Dose rates are defined as the total expected exposure to workers during ISFSI-related operations, or to members of the public who are assumed to be at the closest boundary of the controlled area (at least 100 meters from the storage cask). The NRC has accepted a range of surface dose rates, but doses calculated for workers and the public must comply with the criteria in 10 CFR Parts 20 and 72.

10 CFR Part 72 does not establish specific cask dose rate limits. Cask dose rates from 20 to 400 mrem/hour have been accepted in previous Part 72 evaluations. Acceptable dose rates depend on a number of factors, such as the geometry of the storage array, the time workers will routinely spend in the storage array for activities like monitoring or maintenance, the proximity to other areas frequently occupied by workers, and the proximity to the controlled area boundary or other public access areas.

Review the design criteria presented in Section 2 of the applicant's safety analysis report (SAR), as well as any additional shielding-related criteria. Consider the proximity of the storage array to equipment that must be monitored or serviced frequently. Note that *high dose rates at the cask top or bottom may be acceptable if these areas are not routinely occupied during storage operations and if the expected exposure during cask transfer operations is controlled.* However, remember to evaluate the dose rates at the side of the same cask to ensure that ALARA principles are either engineered into the design or evoked by specific operating procedures.

b. Design Features

Review the general description of the cask presented in SAR Section 1, as well as any additional information provided in SAR Section 5. All drawings, figures, and tables describing shielding features must be sufficiently detailed to support an in-depth staff evaluation.

2. Radiation Source Definition

Examine the description of the design-basis fuel in SAR Section 2 to verify that the applicant calculated the source term on the basis of the fuel that will actually provide the bounding source. The SAR should examine all fuel designs and burnup conditions for which the cask system is to be certified to ensure that the bounding fuel type and values are used. The applicant should devote particular attention to the enrichment, burnup, and cooling times. Generally, the specifications in SAR Section 2 will indicate the maximum fuel enrichment used in the criticality analysis. For shielding evaluations, however, the neutron source term increases considerably with decreasing initial enrichment and constant burnup. Consequently, the SAR may either specify the minimum initial enrichment or establish the specific source terms as operating controls and limits for cask use.

Generally, the applicant will determine the source terms using ORIGEN-S[3] (e.g., as a SAS2 sequence of SCALE), ORIGEN2[4], or the U.S. Department of Energy (DOE) Characteristics Data Base[5]. Although the latter two are easy to use, both have energy group structure limitations, as discussed below. If the applicant has used ORIGEN2, verify that the chosen cross-section library is appropriate for the fuel being considered. Many libraries are not appropriate for a burnup that exceeds 33,000 MWd/MTU.

a. Gamma Source

Verify that the applicant specified gamma source terms as a function of energy for both the spent fuel and activated hardware. If the energy group structure from the source term calculation differs from that of the cross-section set of the shielding calculation, the applicant may need to regroup the photons. Regrouping can be accomplished by using the nuclide activities from the source term calculation as input to a simple decay code with a variable group structure. Some applicants will merely interpolate from one structure to the other. In general, only gammas with energies from approximately 0.8 to 2.5 Mev will contribute significantly to the dose rate through typical types of shielding; thus, regrouping outside this range is of little consequence. Pay attention to whether the source terms are specified per assembly, per total assemblies, or per metric ton, and ensure that the total source is correctly used in the shielding evaluation.

Determining source terms for fuel assembly hardware is generally not as straightforward as for the spent fuel, especially if the applicant uses one of the ORIGEN codes. The effort devoted to reviewing this calculation should be appropriate to the contribution of these terms to the dose rates presented in the shielding evaluation. Also, note whether the cask is intended to contain special hardware, such as control assemblies or shrouds, and ensure that source terms from these components are included if applicable.

Depending on the cask design, neutron interactions may result in the production of energetic gammas being produced near the cask surface. If this source term was not treated by the shielding analysis code, verify that it has been determined by other appropriate means.

As part of the source term determination, the applicant must calculate the quantities of certain nuclides (e.g., Kr^{85}, H^3, and I^{129}) for use in analyzing doses from the release of radioactive material during postulated accidents in later sections of the SAR. If these calculations are presented in the shielding evaluation section, they should be reviewed at this time. Often, the applicant will tabulate all nuclides that are important to the direct radiation dose rate. This information can be used to resolve differences that may exist between the source terms derived by the applicant and those derived by the NRC staff reviewer.

b. Neutron Source

Verify that the neutron source term is expressed as a function of energy. The neutron source will generally result from both spontaneous fission and alpha-n reactions in the fuel. Depending on the method used to determine these source terms, the applicant may need to independently determine the energy group structure. This analysis is often accomplished by selecting the nuclide with the largest contribution to spontaneous fission (e.g., Cm^{244}) and using that spectrum for all neutrons, since the contribution from alpha-neutron reactions is generally small.

3. Shielding Model Specification

Verify that the SAR adequately describes the models that the applicant used in the shielding evaluation for storage under both normal and accident conditions. For example, if the cask has an external neutron shield, determine whether it would be damaged by a tipover accident or degraded in a fire. Coordinate this analysis with the structural and thermal reviews, as appropriate. Confirm that the shielding assumptions made in dose rate calculations for both occupational workers and the public are consistent with the design criteria and design drawings.

a. Configuration of the Shielding and Source

Examine the sketches or figures that indicate how the shielding is modeled. Verify that the model dimensions and materials are consistent with those specified in the cask drawings presented in SAR Section 1. Ensure that voids, streaming paths, and irregular geometries are accounted for or otherwise treated in a conservative manner. In addition, the SAR must clearly state the differences, if any, between normal storage conditions and accident conditions.

Verify that the applicant properly modeled the source term locations for both spent fuel and structural support regions. In some cases, the fuel and basket materials may be homogenized within the fuel region to facilitate the shielding calculations. Watch for cases when homogenization may not be appropriate. For example, homogenization should not be used in neutron dose calculations when significant neutron multiplication can result from moderated neutrons (i.e., when significant amounts of moderating materials are present). Similarly, homogenization should not be used in configurations where significant radiation streaming can occur between the basket components.

Because of a cosine burnup profile, a uniform source distribution is generally conservative for the top and bottom, but not for the axial center. If axial peaking appears to be significant, verify that the applicant has appropriately accounted for the condition. In addition, the structural support regions (e.g., top and bottom end pieces and plenum) of the assembly should be correctly positioned relative to the spent fuel. These support regions may be individually homogenized with the basket materials. Generally, however, at least three source regions (i.e., fuel and top/bottom assembly hardware) are necessary.

Verify that the SAR shows or adequately describes the locations selected for the various dose calculations. Ensure that these dose points are representative of all locations relevant to radiation protection issues. Pay particular attention to dose rates from streaming paths to which occupational workers would be exposed (e.g., at vent/drain port covers, lid bolts, etc.). Also, note the shielding end points, such as lead in the cask wall, in relation to the assembly hardware. See Section IV.4.c, below, for additional information regarding the selection of locations for dose calculations.

b. Material Properties

Verify that the SAR provides information concerning compositions and densities for all materials used in the calculational model. For nonstandard materials (such as neutron shields), SAR Section 9 must also reference the source of the data and indicate validation criteria. Many shielding codes allow the densities to be input directly in g/cm^3. If input is required in atoms/barn-cm, pay particular attention to the conversion.

Confirm that temperature-sensitive shielding materials will not be subject to temperatures at or above their design limitations during either normal or accident conditions. Determine whether the applicant properly examined the potential for shielding material to experience changes in material densities at temperature extremes. (For example, elevated temperatures may reduce hydrogen content through loss of bound or free water in concrete or other hydrogenous shielding materials.)

4. Shielding Analyses

a. Computer Programs

Examine the computer program(s) used for the shielding analysis. These codes may include Monte Carlo, deterministic transport, or point-kernel techniques for problem solution. Some shielding codes available from the Radiation Safety Information Center[a] are listed below:

- TORT\DORT (three- and two-dimensional discrete-ordinate neutron/photon transport codes)

---

[a] Radiation Shielding Information Center, Oak Ridge National Laboratory, P.O. Box 2008, Oak Ridge, Tennessee, 37831-6362.

- ONEDANT/TWODANT (one- and two-dimensional multigroup discrete-ordinate transport codes)

- MCNP (Monte Carlo n-particle transport code)

- ANISN (one-dimensional neutron attenuation code)

- SKYSHINE (air-scattering code)

- MORSE (Monte Carlo multigroup three-dimensional neutron and gamma transport computer code)

- QAD-CGGP (three-dimensional point kernel gamma transport shielding computer code)

- SCALE (a modular code system for performing standardized computer analyses for licensing evaluation)

The above computer programs are recognized and widely known in shielding analysis. However, their use does not constitute generic NRC approval, and the reviewer is cautioned that these codes can produce errors when used incorrectly. The applicant should have design control measures for ensuring the quality of computer programs.

A valuable primer on shielding codes and analysis techniques has been published by Oak Ridge National Laboratory[6].

For each program, verify the following information to demonstrate its applicability and validity:

i. The author, source, dated version, and facility.

ii. A description, and the extent and limitation of its application.

iii. The computer program solutions to a series of test problems, demonstrating substantial similarity to solutions obtained from any one of the following sources:
  (a) hand calculations
  (b) analytical results published in the literature
  (c) acceptable experimental tests
  (d) a similar program
  (e) benchmark problems

In addition, verify that the applicant has prepared a summary comparison of the solutions obtained using sources (a) through (e), in either graphical or numeric form. These solutions may be referenced, and need not be submitted in the SAR, provided that the references are widely publicly available or have previously been submitted to the NRC and the information submitted under items I and ii remains unchanged.

Review the submitted computer solutions to the test problems, and compare them with the test solutions. Satisfactory agreement of computer and test solutions and/or resolution of deviations provides verification of the quality and adequacy of the computer programs to perform the calculations for which they were designed. Identify any deviations that have not previously been justified to the staff's satisfaction, and transmit the finding to the applicant with a request for additional technical justification regarding application of the code.

Determine whether the number of dimensions of the code is appropriate for the dose rates being calculated. Generally, at least a two-dimensional calculation is necessary. One-dimensional codes provide little information about off-axis locations and streaming paths that may be significant for determining occupational exposure. Even for dose rates at the end of the cask, one-dimensional codes require a buckling correction that is not particularly straightforward, since merely using the cask cavity diameter may underestimate the dose rate.

The SAR should include a representative computer code input file. As discussed in Section V.4.d below, if the reviewer is familiar with the code used in the SAR analysis, examining the input file can significantly expedite the review.

Verify that the information from the shielding model is properly entered into the code. Also, verify that the cross-section library used by the code is appropriate for use in analysis of cask shielding problems. If the applicant has not independently determined a source term for neutron-induced gamma radiation or subcritical multiplication of neutrons, ensure that a coupled cross-section set was used and that the applicant executed the code in a manner that accounts for these secondary source terms.

b. Flux-to-Dose-Rate Conversion

The shielding analysis code may perform flux-to-dose-rate conversion using its own data library. For the conversions, the NRC accepts the use of American National Standards Institute/American Nuclear Society (ANSI/ANS) Standard 6.1.1-1977[7]. The 10 CFR Part 20 radiation protection requirements are based on fluence-to-dose conversions that are essentially the same as those defined by ANSI/ANS 6.1.1-1977, and are conservative relative to those of ANSI/ANS 6.1.1-1991. Neutron dose rates determined on the basis of conversions performed according to ANSI/ANS 6.1.1-1991 may be significantly lower than those determined on the basis of 10 CFR Part 20 or ANSI/ANS 6.1.1-1977.

c. Dose Rates

On the basis of experience, comparison to similar systems, or scoping calculations, make an initial assessment of whether the dose rates appear reasonable and whether their variation with location is consistent with the geometry and shielding characteristics of the cask system. The following guidance pertains to the selection of points at which the dose rates should be calculated.

For normal conditions, the SAR should indicate the dose rate at all locations accessible to occupational personnel during cask loading, transport to the ISFSI, and maintenance and surveillance operations. Generally these locations include points at or near various cask components and in the immediate vicinity of the cask.

Appropriately detailed calculations are necessary to show that the radiation shielding features are sufficient to meet the requirements of 10 CFR 72.104 and 72.106. These calculations will need to assume a typical storage arrangement for the casks. Later, when a particular site is selected, calculations will be needed to show ultimate compliance of the spent fuel system. Site configurations not enveloped by the typical site layout assumed in the SAR must be treated in the written evaluations required under 72.212(b)(2) and (3) before the cask system is used. In addition, the SAR should determine the degree to which the normal condition dose rates could change for the identified off-normal conditions. The need for additional calculations should be indicated in the SER and in the conditions set forth in the Certificate of Compliance.

The NRC has previously accepted a calculated dose rate of 0.25 mSv/yr (25 mrem/yr) at the ISFSI controlled area boundary as sufficient evidence that the limit for exposure of the public will not be exceeded under normal conditions. The applicant should provide a discussion that would help determine whether a potential ISFSI would be within the dose rate envelope. These could involve identifying the minimum controlled area dimensions to ensure that the 0.25 mSv/yr dose is not exceeded. Alternatively, the presentation could provide the maximum number of casks that could be stored in an ISFSI, with the minimum distance of 100 meters between the stored fuel and the controlled area boundary (10 CFR Part 72.106(b)), or a suggestion that the licensee install berms, to meet the criterion of 0.25 mSv/yr (25 mrem/yr).

To demonstrate applicant compliance with these requirements, the NRC staff has accepted calculations in the SAR showing a dose rate less than 0.25 mSv/yr (25 mrem/yr) from one cask (or a representative array of casks) at an assumed distance to the controlled area boundary. Such calculations, in practice, can give only a general assessment of the proposed cask system. In addition to unknown information about the ISFSI itself, the implied assumption that an individual would be at the controlled area boundary for 8760 hours (the entire year) is very conservative.

If the above dose rate criteria are satisfied, NRC accepts that the direct-dose regulatory requirements can also be satisfied, although the exact details needed to comply with these limitations will vary from site to site. Therefore, the SAR needs to address such requirements only in general terms. Detailed calculations need not be presented if SAR Section 12, "Operating Controls and Limits," assigns ultimate compliance responsibilities to the site licensee.

In addition, the applicant should calculate the dose rate at one meter from the cask surface for accident conditions. The model used for these calculations must be consistent with the expected condition of the cask after an accident or natural phenomenon event.

d. Confirmatory Calculations

Independently evaluate the dose rates in the vicinity of the cask for both normal and accident conditions. In determining the level of effort appropriate for these calculations, consider the following factors:

- the degree of sophistication and margin in the SAR analysis

- a comparison of SAR dose rates with those of similar casks that have previously been reviewed, if applicable

- the typical variation in dose rates expected between different codes and cross-section sets

- the fact that actual dose rates will be monitored and limited by the requirements of 10 CFR Part 20

- the restrictions that can be placed on ISFSI operations affecting measured dose rates, as documented in SER Section 12, the site-specific license, or the Certificate of Compliance

- the applicant's experience in using the methods and codes in previous ISFSI submittals

- use of new, or previously reviewed, methods or codes

- inclusion in the design of any significant departures from previous cask system designs (e.g., unusual shield geometry, new types of materials, or different source terms)

At a minimum, the review should include examination of the applicant's input to the computer program used for the shielding analysis. Verify use of proper dimensions, material properties, and an appropriate cross-section set. In addition, independently evaluate the use of gamma and neutron source terms.

If a more detailed review is required, independently evaluate the dose rates to ensure that the SAR results are reasonable and conservative. As previously noted, the use of a simple code for neutron calculations is often not appropriate. An extensive evaluation is necessary if major errors are suspected. To the degree possible, the use of a different shielding code with a different analytical technique and cross-section set from that of the SAR analysis will provide a more independent evaluation.

A good reference regarding the treatment of uncertainty in thick-shielded cask analyses has been published by the Electric Power Research Institute[b].

5. Supplemental Information

Supplemental information can include copies of applicable references (especially if a reference is not generally available to the reviewer), computer code descriptions, input and output files, and any other information that the applicant has deemed necessary. Likewise, the reviewer should request any additional information needed to complete the review process.

VI. Evaluation Findings

Review the 10 CFR Part 72 acceptance criteria and provide a summary statement for each. These statements should be similar to the following model:

Section(s) _____ of the SAR describe(s) shielding structures, systems, and components (SSCs) important to safety in sufficient detail to allow evaluation of their effectiveness.

---

[b]   B.L. Broadhead, *et. al.*, Electric Power Research Institute, "Evaluation of Shielding Analysis Methods in Spent Fuel Cask Environments," EPRI TR-104329, Palo Alto, California, May 1995.

Section(s) _____ of the SAR evaluates these shielding SSCs important to safety with the objective of assessing the impact on health and safety resulting from operation of the independent spent fuel storage installation (ISFSI).

The staff concludes that the design of the shielding system of the [cask designation] is in compliance with 10 CFR Part 72 and that the applicable design and acceptance criteria including 10 CFR Part 20 have been satisfied. The evaluation of the shielding system design provides reasonable assurance that the [cask designation] will allow safe storage of spent fuel. This finding is reached on the basis of a review that considered the regulation itself, appropriate regulatory guides, applicable codes and standards, and accepted engineering practices.

VII. References

1.   *U.S. Code of Federal Regulations*, "Licensing Requirements for the Independent Storage of Spent Nuclear Fuel and High-level Radioactive Waste," Part 72, Title 10, "Energy."

2.   *U.S. Code of Federal Regulations*, Part 20, "Standards for Protection Against Radiation," Title 10, "Energy."

3.   L.M. Petrie, *et al.*, Oak Ridge National Laboratory, "SCALE: A Modular Code System for Performing Standardized Computer Analyses for Licensing Evaluation," NUREG/CR-0200, Vol. 1–4, Rev. 4, April 1995.

4.   Oak Ridge National Laboratory, "ORIGEN2.1: Isotope Generation and Depletion Code—Matrix Exponential Method," 1991.

5.   TRW Environmental Safety Systems, Inc., "DOE — OCRWM Characteristics Database," DOE/RW-0184-R1.

6.   C.V. Parks, *et. al.*, Oak Ridge National Laboratory, "Assessment of Shielding Analysis Methods, Codes, and Data for Spent Fuel Transport/Storage Applications," ORNL/CSD/TM-246, July 1988.

7.   American National Standards Institute/American Nuclear Society (ANSI/ANS) Standard 6.1.1-1977.

6.0 CRITICALITY EVALUATION

I. Review Objective

The criticality review ensures that spent fuel remains subcritical under normal, off-normal, and accident conditions involving handling, packaging, transfer, and storage.

II. Areas of Review

This portion of the dry cask storage system (DCSS) review evaluates the criticality design and analysis related to spent fuel handling, packaging, transfer, and storage procedures for normal, off-normal, and accident conditions. Consequently, this chapter of the DCSS Standard Review Plan (SRP) provides guidance for use in conducting a comprehensive criticality evaluation that *may* encompass any or all of the following areas of review:

1. criticality design criteria and features
2. fuel specification
3. model specification
   a. configuration
   b. material properties
4. criticality analysis
   a. computer programs
   b. multiplication factor
   c. benchmark comparisons
5. supplemental information

III. Regulatory Requirements

Spent fuel storage systems must be designed to remain subcritical unless at least two unlikely independent events occur. Moreover, the spent fuel cask must be designed to remain subcritical under all credible conditions. Regulations specific to nuclear criticality safety of the cask system are specified in 10 CFR 72.124 and 72.236(c). Other pertinent regulations include 10 CFR 72.24(c)(3), 72.24(d), and 72.236(g). Normal and accident conditions to be considered are also identified in 10 CFR Part 72.

IV. Acceptance Criteria

In general, the DCSS criticality evaluation seeks to ensure that the given design fulfills the following acceptance criteria:

1. The multiplication factor (k_{eff}), including all biases and uncertainties at a 95-percent confidence level, should not exceed 0.95 under all credible normal, off-normal, and accident conditions.

2. At least two unlikely, independent, and concurrent or sequential changes to the conditions essential to criticality safety, under normal, off-normal, and accident conditions, should occur before an accidental criticality is deemed to be possible.

3. When practicable, criticality safety of the design should be established on the basis of favorable geometry, permanent fixed neutron-absorbing materials (poisons), or both. Where solid neutron-absorbing materials are used, the design should provide for a positive means to verify their continued efficacy during the storage period.

4. Criticality safety of the cask system should not rely on use of the following credits:

   a. burnup of the fuel
   b. fuel-related burnable neutron absorbers

c. more than 75 percent for fixed neutron absorbers[a] when subject to standard acceptance tests.

V. Review Procedures

Review the criticality design features and criteria in SAR Chapters 1 and 2. Also review SAR Chapter 6 for any additional details concerning criticality design features and criteria. Assess the bounding specifications for the spent fuel. Examine the models used by the applicant in the criticality analyses. Verify that the applicant has addressed criticality safety considerations under normal, off-normal, and accident conditions. Verify that the cask system design complies with 10 CFR Part 72. In addition, verify that the criticality calculations determine the highest k_{eff} that might occur under all loading states under normal, off-normal, and accident conditions involving handling, packaging, transfer, or storage. To the extent practicable, use independent methods to perform any k_{eff} calculations to evaluate the applicant's design.

1. Criticality Design Criteria and Features

Review the principal criticality design criteria presented in SAR Chapter 2, as well as any related detail provided in SAR Chapter 6. Also review the general cask description presented in SAR Chapter 1 and any related information provided in Chapter 6. Verify that the information in Chapter 6 is consistent with the information in Chapters 1 and 2. Also, verify that all drawings, figures, and tables are sufficiently detailed to support in-depth staff evaluation.

In addition to the general dimensions of the cask components and spacing of fuel assemblies in the basket, the criticality design often relies on neutron poisons. These may be in the form of fixed poisons in the basket structure and/or soluble poisons in the water of the spent fuel pool. The NRC staff accepts the use of borated water as a means of criticality control if the applicant specifies a minimum boron content, and strict controls are established to ensure that the minimum required boron concentration is maintained, which in turn becomes an operating control and limit in SAR Chapter 12. These operating controls should also be discussed in the SER. If borated water is used for criticality control, administrative controls and/or design features should be implemented to ensure that accidental flooding with unborated water cannot occur, or the criticality evaluation should consider accidental flooding with unborated water. If the cask is also intended for transport, borated water cannot be relied upon for criticality control.

2. Fuel Specification

Review the specifications for the ranges or types of spent fuel that will be stored in the cask as presented in SAR Sections 1 and 2, as well as any related information provided in SAR Sections 6. Verify that the spent fuel specifications given in Section 6 are consistent with, or bounded by, the specifications given in Section 1 and 2.

Of primary interest is the type of fuel assemblies and maximum fuel enrichment, which should be specified and used in the criticality calculations. Some boiling water reactors (BWR) use multiple fuel pin enrichments, in which case, the criticality calculations should use the maximum fuel pin enrichment present. Depending upon the fuel design, an applicant may propose use of assembly averaged, or lattice averaged enrichments. This may be acceptable if the applicant can demonstrate that any averaging techniques are technically defensible and, for the criticality calculation, produce conservative results. Because of the natural uranium blankets present in many BWR designs, use of an assembly-averaged enrichment is not normally considered appropriate or conservative for BWR fuel.

Although the burnup of the fuel affects its reactivity, the NRC staff does not currently allow credit for burnup, either in depleting the quantity of fissile nuclides or in producing fission product poisons for spent fuel storage or transport casks. Specifications for the fuel that will be stored in the cask should be included in Section 12 of both the SAR and SER and should also be explicitly listed in the Certificate of Compliance.

The fresh fuel assumption should be used in the criticality analyses; therefore, inadvertent loading of the cask with unirradiated fuel is not a major concern. Nonetheless, detailed loading procedures may need to

---

[a] For greater credit allowance, special, comprehensive fabrication tests capable of verifying the presence and uniformity of the neutron absorber are needed.

include steps to prevent misloading if fuel exceeding the design basis for the DCSS is present in the pool at the time of loading.

Because casks are typically designed to store many types and configurations of fuel assemblies, the applicant should demonstrate that criticality requirements are satisfied for the most reactive case. A determination of which fuel is bounding in a criticality analysis depends on many factors and usually requires examination of several types of fuel assemblies and compositions. The design-basis fuel has often been the Westinghouse 17x17 optimized fuel assembly (OFA); however, this will not be the case for all cask designs because of cask-specific effects on reactivity. Therefore, the applicant should demonstrate and reviewers should verify, that the fuel assembly used as the design basis is the most reactive for the specific cask design. Chapter 12 of both the SAR and SER should either clearly indicate the design-basis assemblies or reference the SAR chapter in which they are identified.

Determine if the applicant has included any specifications regarding the fuel condition. To date, casks have not typically been intended to store fuel that is significantly damaged or has a gross cladding defect. Consequently, the criticality analyses have generally specified that any damaged fuel rods should be replaced with dummy rods that can displace an equal amount of water as the original rods. If invoked by the applicant, these requirements, should be included as operating controls and limits and discussed in SAR Chapter 12.

3. Model Specification

When manufacturing and fabrication tolerances are specified, verify that the applicant assumed the most conservative value within the range of acceptable values.

a. Configuration

Verify that the model used in the criticality evaluation is adequately described for normal, off-normal, and accident conditions. Coordinate with the structural reviewer to understand any damage that could result from accident or natural phenomena events.

Examine the sketches or figures of the model used for criticality calculations. Verify that the dimensions and materials of the model are consistent with those in the drawings of the actual cask. Differences between the actual cask configuration and the models should be identified, and the models should be shown to be conservative. Substitution of ordinary water for end sections and support structures of the fuel is a common and conservative practice in criticality analysis; however, substitution of borated water is non-conservative. Tolerances for poison material dimensions and/or concentrations should be defined, and the most reactive conditions should be used in the criticality analysis. In addition, the analysis should identify all important design conditions and then address these conditions for normal, off-normal, and accident conditions.

Verify that the applicant has considered deviations from nominal design configurations. The evaluation of k_{eff} should not be limited to a model in which all of the fuel bundles are neatly centered in each basket compartment with the center line of the basket coincident with the center line of the cask. For example, a cask with steel confinement and lead shielding may have a higher k_{eff} when the basket and fuel assemblies are positioned as close as possible to the lead.

In addition to a fully flooded cask, the SAR should address configurations in which the cask is partially filled with water (borated, if applicable) and the remainder of the cask is filled with steam consisting of ordinary water at partial density. These configurations are considered to be representative of loading and unloading operations in the spent fuel pool. The SAR should also consider the possibility of preferential or uneven flooding within the cask, if such a scenario is credible for the given cask design (e.g., because of blockage in small flow or drain paths). In particular, watch for situations where there is water in the fuel regions but not in the flux traps, if applicable. Cask designs for which this type of flooding is credible are generally unacceptable. The SAR should also consider flooding in the fuel rod pellet-to-clad gap regions. Above all, the analysis must demonstrate that the cask remains subcritical for all credible conditions of moderation.

Examine whether the applicant has prepared a heterogeneous model of each fuel rod or has homogenized the entire fuel assembly. With current computational capabilities, homogenization is now an uncommon practice and should generally be avoided. If such homogenization is used, however, the applicant should clearly demonstrate that it has been treated conservatively. As a minimum, the applicant should calculate the k_{eff} of one assembly and several critical benchmark experiments (see Section V(4)(c) of this chapter)

Criticality Evaluation

using both homogeneous and heterogeneous models. The applicant should then compare the results of these calculations.

b. Material Properties

Verify that the compositions and densities are provided for all materials used in the calculational model. The applicant should also cite the source of all materials data, particularly the data for fuel and poison materials. Ensure that the applicant addressed the validation of the poison concentration in the acceptance testing discussion in SAR Chapter 9. Many criticality codes will allow the densities to be input directly in units of g/cm^3. If input is in units of atoms/barn-cm, pay particular attention to the conversion.

Among other specifications, 10 CFR Part 72 requires that when solid neutron-absorbing materials are used, a positive means to verify their continued efficacy should be provided. Continued efficacy can be demonstrated in the following ways:

- Require acceptance testing of the poisons during fabrication (specified in SAR Chapter 9),

- Show that the neutron flux from the irradiated fuel results in a negligible depletion of poison material over the storage period, and

- Assess the structural integrity and potential for poison material degradation during storage.

If continued efficacy can be demonstrated by design and material properties, a surveillance or monitoring program to "verify" continued efficacy of solid neutron absorbers may not be necessary. The neutron flux used for this analysis should be the maximum that may be produced by feasible loadings of irradiated or unirradiated fuel.

Determine whether the applicant has chosen an acceptable set of cross-sections. Cross-sections may be distributed with the criticality computer codes or developed independently from another source. The applicant should provide or reference the source of cross-section data, as well as the method used to obtain the actual data employed in the criticality analysis. For multigroup calculations, the neutron flux spectrum used to construct the group cross-sections should be similar to that of the cask.

4. Criticality Analysis

a. Computer Programs

Both Monte Carlo and deterministic computer codes may be used for criticality calculations. Monte Carlo codes are generally better suited to three-dimensional geometry and, therefore, are more widely used to evaluate spent fuel cask designs. The two most frequently used Monte Carlo codes are SCALE/KENO[1] and MCNP[2]. KENO is a multigroup code that is part of the SCALE sequence, while MCNP permits the use of continuous cross-sections.

If a multigroup treatment is used, ensure that the applicant has appropriately considered the neutron spectrum of the cask. In addition to selecting a cross-section set collapsed with an appropriate flux spectrum, a more detailed processing of the energy-group cross-sections is also required to properly account for resonance absorption and self-shielding. The use of KENO as part of the SCALE sequence will directly enable such processing. Some cross-section sets include data for fissile and fertile nuclides (based on a potential scattering cross-section, s$_p$) that can be input by the user. If the applicant has used a stand-alone version of KENO, ensure that potential scattering has been properly considered. Furthermore, the "working-format" library, once (commonly) distributed with SCALE/KENO to facilitate calculations of the code-manual's sample problems, is not intended for criticality calculations of actual systems. In 1991, the staff provided information concerning cross-section problems to all ISFSI licensees, applicants, and dry storage vendors[3].

For analyses of a cask model with separate regions of water and steam, the use of a multigroup cross-section set raises additional concerns. Verify that the applicant has addressed the differences of the flux spectra in the two regions. If the results of these calculations indicate that k$_{eff}$ is close to 0.95, additional independent calculations using a different code and/or cross-section library may be helpful. Reviewers should also closely examine the applicant's benchmark analysis, in order to verify the applicability of critical experiments considered.

b. Multiplication Factor

Examine the results and discussion of the k_{eff} calculations for the storage cask. Determine if variations in the results caused by different models and sensitivity analyses can be explained and appear reasonable.

For Monte Carlo calculations, assess if the number of neutron histories and convergence criteria are appropriate. As the number of neutron histories increases, the mean value for k_{eff} should approach some fixed value, and the standard deviation associated with each mean value should decrease. Depending on the code used by the applicant, a number of diagnostic calculations are generally available to demonstrate adequate convergence and statistical variation. For deterministic codes, a convergence limit is often prescribed in the input. The selection of a proper convergence limit and the achievement of this limit should be described and demonstrated.

Because of the importance and complexity of the criticality evaluation, independent calculations should be performed to ensure that the most reactive conditions have been addressed and that the reported k_{eff} is conservative. In deciding the level of effort necessary to perform independent confirmatory calculations, the reviewer should consider the following three factors: (1) the calculational method (computer code) used by the applicant; (2) the degree of conservatism in the applicant's assumptions and analyses; and (3) how large a margin exists between the calculated result and the acceptance criterion of $K_{eff} \leq 0.95$. As with any design and review, a small margin below the acceptance criterion and/or a small degree of conservatism necessitate a more extensive analysis.

As the reviewer, develop a model that is independent of the applicant's. If the reported k_{eff} for the worst case is substantially lower than the acceptance criterion of 0.95, a simple model known to produce very conservative results may be all that is necessary for the independent calculations.

If possible and appropriate, perform the independent calculations with a computer code different from that used by the applicant. Likewise, use of a different cross-section set can provide a more independent confirmation.

Although a k_{eff} of 0.95 or lower meets the acceptance criterion, reviewers should watch for design features or content specifications where small changes could result in large changes in the value of k_{eff}. When the value of k_{eff} is highly sensitive to system parameters that could vary, the acceptable k_{eff} limit should be reduced below 0.95. When establishing a k_{eff} limit below 0.95, reviewers should consider the degree of sensitivity to system parameter changes, and the likelihood and extent of potential parameter variations.

c. Benchmark Comparisons

Computer codes for criticality calculations should be benchmarked against critical experiments. A thorough comparison provides justification for the validity of the computer code, its use for a specific hardware configuration, the neutron cross-sections used in the analysis, and consistency in modeling by the analyst. (Using the benchmark results for calculations performed by another analyst does not address this last issue.) The calculated k_{eff} of the cask should then be adjusted to include the appropriate biases and uncertainties from the benchmark calculations.

Examine the general description of the benchmark comparisons. Verify that the analysis of the experiments used the same computer code, hardware, and cross-section data as those used to calculate the k_{eff} values for the cask.

Reviewers should also closely examine the applicant's benchmark analysis to determine whether the benchmark experiments are relevant to the actual cask design. No critical benchmark experiment will precisely match the fissile material, moderation, neutron poisoning, and configuration in the actual cask. However, the applicant can perform a proper benchmark analysis by selecting experiments that adequately represent cask and fuel features and parameters that are important to reactivity. Key features and parameters that should be considered in selecting appropriate critical experiments include the type of fuel, enrichment, hydrogen to uranium (H/U) ratio (dependent largely on rod diameter and pitch), reflector material, neutron energy spectrum, and poisoning material and placement. The applicant should justify, and reviewers should verify, the suitability of the critical experiments chosen to benchmark the criticality code and calculations. UCID-21830[4] provides information on benchmark experiments that may apply to the cask being analyzed.

Reviewers need to assess whether the applicant analyzed a sufficient number of appropriate benchmark experiments, and how the results of these benchmark calculations have been converted to a bias for the cask calculations. Simply averaging the biases from a number of benchmark calculations is typically not sufficient, particularly if one benchmark yields results that are significantly different from the others. In addition, benchmark comparisons should be checked for bias trends with respect to parameter variations (such as pitch-to-rod-diameter ratio, assembly separation, reflector material, neutron absorber material, etc.). Ref. 4 provides some guidance, but other methods have also been considered appropriate.

For Monte Carlo codes, the statistical uncertainties of both benchmark and cask calculations also need to be addressed. The uncertainties should be applied to at least the 95-percent confidence level. As a general rule, if the acceptability of the result depends on these rather small differences, reviewers should question the overall degree of conservatism of the calculations. Considering the current availability of computer resources, a sufficient number of neutron histories can readily be used so that the treatment of these uncertainties should not significantly affect the results.

Verify that only biases that increase k_{eff} have been applied. For example, if the benchmark calculation for a critical experiment results in a neutron multiplication that is greater than unity, it should not be used in a manner that would reduce the k_{eff} calculated for the cask. Only corrections that increase k_{eff} should be applied to preserve conservatism.

Reviewers may have already performed a number of benchmark calculations applicable to storage casks and may have a reasonable estimation of the bias to be applied to the independent calculation of the cask. If such is not the case, or if the acceptability depends on small bias differences, reviewers again need to determine whether sufficient conservatism has been applied to the calculations.

5. Supplemental Information

Ensure that all supportive information or documentation is provided. This would include, but not be limited to, justification of assumptions or analytical procedures, test results, photographs, computer program descriptions, input/output, and applicable pages from referenced documents. In addition, the SER should include a list of fuel designs with the acceptable parametric limits, and the maximum enrichments for which the criticality analysis is satisfactory. Reviewers should request any additional information needed to complete the review.

VI. Evaluation Findings

Review the 10 CFR Part 72 acceptance criteria and provide a summary statement for each. These statements should be similar to the following model:

Structures, systems, and components important to criticality safety are described in sufficient detail in Chapters _____ of the SAR to enable an evaluation of their effectiveness.

The _____ cask and its spent fuel transfer systems are designed to be subcritical under all credible conditions.

The criticality design is based on favorable geometry, fixed neutron poisons, and soluble poisons of the spent fuel pool [as applicable]. An appraisal of the fixed neutron poisons has shown that they will remain effective for the 20-year storage period, and there is no credible way to lose it, therefore there is no need to provide a positive means to verify their continued efficacy as required by 10 CFR 72.124(b).

The analysis and evaluation of the criticality design and performance have demonstrated that the cask will enable the storage of spent fuel for a minimum of 20 years with an adequate margin of safety.

The staff concludes that the criticality design features for the [cask designation] are in compliance with 10 CFR Part 72, as exempted [if applicable], and that the applicable design and acceptance criteria have been satisfied. The evaluation of the criticality design provides reasonable assurance that the [cask designation] will allow safe storage of spent fuel. This finding is reached on the basis of a review that considered the regulation itself, appropriate regulatory guides, applicable codes and standards, and accepted engineering practices.

VII. References

1. U.S. Nuclear Regulatory Commission, "SCALE: A Modular Code System for Performing Standardized Computer Analyses for Licensing Evaluation," NUREG/CR-0200, Vol. 1–4, Rev. 4, April 1995.

2. Los Alamos National Laboratory, "MCNP 4A, Monte Carlo N-Particle Transport Code System," December 1993.

3. U.S. Nuclear Regulatory Commission, "Potential Nonconservative Errors in the Working Format Hansen-Roach Cross-Section Set Provided with the KENO and SCALE Codes," Information Notice No. 91-26, April 15, 1991.

4. W.R. Lloyd, Lawrence Livermore National Laboratory, "Determination and Application of Bias Values in the Criticality Evaluation of Storage Cask Designs," UCID-21830, January 1990. (This report contains a substantial bibliography of numerous benchmark experiments and validation testing.)

7.0 CONFINEMENT EVALUATION

I. Review Objective

In this portion of the dry cask storage system (DCSS) review, the NRC evaluates the confinement features and capabilities of the proposed cask system. In conducting this evaluation, the NRC staff seeks to ensure that radiological releases to the environment will be within the limits established by the regulations and that the spent fuel cladding and fuel assemblies will be sufficiently protected during storage against degradation that might otherwise lead to gross ruptures.

II. Areas of Review

This chapter of the DCSS Standard Review Plan (SRP) provides guidance for use in evaluating the design and analysis of the proposed cask confinement system for normal, off-normal, and accident conditions. This evaluation includes a more detailed assessment of the confinement-related design features and criteria initially presented in Sections 1 and 2 of the applicant's safety analysis report (SAR), as well as the proposed confinement monitoring capability, if applicable. In addition, the NRC staff assesses the anticipated releases of radionuclides associated with spent fuel, by independently estimating their leakage to the environment and the subsequent impact on a hypothetical individual located beyond the controlled area boundary.

As prescribed in 10 CFR Part 72, the regulatory requirements for doses at and beyond the controlled area boundary include both the direct dose and that from an estimated release of radionuclides to the atmosphere (based on the tested leaktightness of the confinement). Thus, an overall assessment of the compliance of the proposed DCSS with these regulatory limits is deferred until Chapter 10, "Radiation Protection," of this SRP. In addition, the performance of the cask confinement system under accident conditions, as evaluated in this section, may also be addressed in the overall accident analyses, as discussed in Chapter 11 of this SRP.

As described in Section V, "Review Procedures," a comprehensive confinement evaluation *may* encompass the following areas of review:

1. confinement design characteristics
   a. design criteria
   b. design features
2. confinement monitoring capability
3. nuclides with potential for release
4. confinement analyses
   a. normal conditions
   b. leakage of one seal
   c. accident conditions and natural phenomenon events
5. supplemental information

III. Regulatory Requirements

1. Description of Structures, Systems, and Components Important to Safety

The SAR must describe the confinement structures, systems, and components (SSCs) important to safety in sufficient detail to facilitate evaluation of their effectiveness. [10 CFR 72.24(c)(3) and 10 CFR 72.24(l)]

2. Protection of Spent Fuel Cladding

The design must adequately protect the spent fuel cladding against degradation that might otherwise lead to gross ruptures during storage, or the fuel must be confined through other means such that fuel degradation during storage will not pose operational safety problems with respect to removal of the fuel from storage. [10 CFR 72.122(h)(1)]

3. Redundant Sealing

The cask design must provide redundant sealing of the confinement boundary. [10 CFR 72.236(e)]

4. Monitoring of Confinement System

Storage confinement systems must allow continuous monitoring, such that the licensee will be able to determine when to take corrective action to maintain safe storage conditions. [10 CFR 72.122(h)(4) and 10 CFR 72.128(a)(1)]

5. Instrumentation

The design must provide instrumentation and controls to monitor systems that are important to safety over anticipated ranges for normal and off-normal operation. In addition, the applicant must identify those control systems that must remain operational under accident conditions. [10 CFR 72.122(i)]

6. Release of Nuclides to the Environment

The applicant must estimate the quantity of radionuclides expected to be released annually to the environment. [10 CFR 72.24(l)(1)]

7. Evaluation of Confinement System

The applicant must evaluate the cask and its systems important to safety, using appropriate tests or other means acceptable to the Commission, to demonstrate that they will reasonably maintain confinement of radioactive material under normal, off-normal, and credible accident conditions. [10 CFR 72.236(l) and 10 CFR 72.24(d)]

In addition, SSCs important to safety must be designed to withstand the effects of credible accidents and severe natural phenomena without impairing their capability to perform safety functions. [10 CFR 72.122(b)]

8. Annual Dose Limit in Effluents and Direct Radiation from an Independent Spent Fuel Storage Installation (ISFSI)

During normal operations and anticipated occurrences, the annual dose equivalent to any real individual who is located beyond the controlled area must not exceed 25 mrem to the whole body, 75 mrem to the thyroid, and 25 mrem to any other organ. [10 CFR 72.104(a)]

IV. Acceptance Criteria

In general, DCSS confinement evaluation seeks to ensure that the proposed design fulfills the following acceptance criteria, which the NRC staff considers to be minimally acceptable to meet the confinement requirements of 10 CFR Part 72:

1. The cask design must provide redundant sealing of the confinement boundary sealing surface. Typically, this means that field closures of the confinement boundary must either have double seal welds or double metallic O-ring seals.

2. The confinement design must be consistent with the regulatory requirements, as well as the applicant's "General Design Criteria" reviewed in Chapter 2 of this SRP. The NRC staff has accepted construction of the primary confinement barrier in conformance with Section III, Subsections NB or NC, of the Boiler and Pressure Vessel (B&PV) Code[1] promulgated by the American Society of Mechanical Engineers (ASME). (This code defines the standards for all aspects of construction, including materials, design, fabrication, examination, testing, inspection, and certification required in the manufacture and installation of components.) In such instances, the staff has relied upon Section III to define the minimum acceptable margin of safety; therefore, the applicant must fully document and completely justify any deviations from the specifications of Section III. In some cases after careful and deliberate consideration, the staff has made exceptions to this requirement.

3. The applicant must specify the maximum allowed leakage rates for the total primary confinement boundary and redundant seals. (Applicants frequently display this information in tabular form, including the leakage rate of each seal.) In addition, the applicant's leakage analysis should be consistent with the principles specified in the "American National Standard for Leakage Tests on

Packages for Shipment of Radioactive Materials" (ANSI N14.5)[2]. Generally, the allowable leakage rate must be evaluated for its radiological consequences and its effect on maintaining the necessary inert atmosphere within the cask.

4. The applicant should describe the proposed monitoring capability and/or surveillance plans for mechanical closure seals. In instances involving welded closures, the staff has previously accepted that no closure monitoring system is required. This practice is consistent with the fact that other welded joints in the confinement system are not monitored. However, the lack of a closure monitoring system has typically been coupled with a periodic surveillance program that would enable the licensee to take timely and appropriate corrective actions to maintain safe storage conditions after closure degradation. The discussion in (a) below taken from chapter 2 of this SRP expands on the requirement for continuous monitoring.

   (a) Continuous Monitoring

   The Office of the General Counsel (OGC) has developed an opinion as to what constitutes "continuous monitoring" as required in 10 CFR Part 72.122(h)(4). The staff, in accordance with that opinion has concluded that both routine surveillance programs and active instrumentation meets the intent of "continuous monitoring." Cask vendors may propose, as part of the SAR, either active instrumentation and/or surveillance to show compliance with 10 CFR Part 72.122(h)(4).

   The reviewer should note that some DCSS designs may contain a component or feature whose continued performance over the licensing period has not been demonstrated to staff with a sufficient level of confidence. Therefore the staff may determine that active monitoring instrumentation is required to provide for the detection of component degradation or failure. This particularly applies to components whose failure immediately affects or threatens public health and safety. In some cases the vendor or staff in order to demonstrate compliance with 10 CFR Part 72.122(h)(4), may propose a technical specification requiring such instrumentation as part of the initial use of a cask system. After initial use, and if warranted and approved by staff, such instrumentation may be discontinued or modified.

5. The cask must provide a non-reactive environment to protect fuel assemblies against fuel cladding degradation, which might otherwise lead to gross rupture.[3] Measures for providing a non-reactive environment within the confinement cask typically include drying, evacuating air and water vapor, and backfilling with a non-reactive cover gas (such as helium). For dry storage conditions, experimental data have not demonstrated an acceptably low oxidation rate for UO_2 spent fuel, over the 20-year licensing period, to permit safe storage in an air atmosphere. Therefore, to reduce the potential for fuel oxidation and subsequent cladding failure, an inert atmosphere (e.g., helium cover gas) has been used for storing UO_2 spent fuel in a dry environment. (See Chapter 8 of this SRP for more detailed information on the cover gas filling process.) Note that other fuel types, such as graphite fuels for the high-temperature gas-cooled reactors (HTGRs), may not exhibit the same oxidation reactions as UO_2 fuels and, therefore, may not require an inert atmosphere. Applicants proposing to use atmospheres other than inert gas should discuss how the fuel and cladding will be protected from oxidation.

V.   Review Procedures

1.   Confinement Design Characteristics

a.   Design Criteria

Review the principal design criteria presented in SAR Section 2, as well as any additional detail provided in SAR Chapter 7.

b.   Design Features

Review the general description of the cask presented in SAR Section 1, as well as any additional information provided in SAR Section 7. All drawings, figures, and tables describing confinement features must be sufficiently detailed to stand alone.

Verify that the applicant has clearly identified the confinement boundaries. This identification should include the confinement vessel; its penetrations, valves, seals, welds, and closure devices; and corresponding information concerning the redundant sealing.

Verify that the design and procedures provide for drying and evacuation of the cask interior as part of the loading operations, and that the design is acceptable for the pressures that may be experienced during these operations.

Verify that, on completion of cask loading, the gas fill of the cask interior is at a pressure level that is expected to maintain a non-reactive environment for at least the 20-year storage life of the cask interior under both normal and off-normal conditions and events. This verification can include pressure testing, seal monitoring, and maintenance for casks with seals that are not welded if these are included in chapter 12 as conditions of use. The NRC has previously accepted specification of an overpressure of approximately 14 kiloPascals (~2 psig) and cask leak testing as conditions of use for satisfying this requirement. In addition, if conditions of use require routine inspection of seals by the pressure testing of the cask interior, the cask fill pressure may be linked to that activity.

Coordinate with the structural reviewer (Chapter 3 of this SRP) to ensure that the applicant has provided proper specifications for all welds and, if applicable, that the bolt torque for closure devices is adequate and properly specified.

If applicable, assess the seals used to provide closure. Because of the performance requirements over the 20-year license period, evaluate the potential for deterioration. The NRC staff has previously accepted only metallic seals for the primary confinement. Coordinate with the thermal reviewers (Chapter 4 of this SRP) to ensure that the operational temperature range for the seals, specified by the manufacturer, will not be exceeded.

2. Confinement Monitoring Capability

The NRC staff has found that casks closed entirely by welding do not require seal monitoring. However, for casks with bolted closures, the staff has found that a seal monitoring system has been needed in order to adequately demonstrate that seals can function and maintain a helium atmosphere in the cask for the 20-year license period. A seal monitoring system combined with periodic surveillance enables the licensee to determine when to take corrective action to maintain safe storage conditions. (Note that some designs may not require an inert atmosphere in the cask. In such designs, a periodic surveillance program to check seal leak tightness may be appropriate.)

Although the details of the monitoring system may vary, the general design approach has been to pressurize the region between the redundant seals, with a non-reactive gas, to a pressure greater than that of the cask cavity and the atmosphere. A decrease in pressure between these seals indicates that the non-reactive gas is leaking either into the cask cavity or out to the atmosphere. (Radioactive gas should not be able to leak to the atmosphere in either case; hence, a faulty seal can be detected without radiological consequence.) Note that the volume between the redundant seals should be pressurized using a *non-reactive* gas, thereby preventing contamination of the interior cover gas.

The monitoring system is generally not important to safety and, as such, is classified as Category B under the guidelines of NUREG/CR-6407[4]. Although its function is to monitor confinement seal integrity, failure of the monitoring system does not result in a release of radioactive material. Consequently, the monitoring system for bolted closures need not be designed to the same requirements as the confinement boundary (i.e., ASME Section III, Subsections NB or NC).

In order to meet confinement boundary design standards, either the entire pressurized portion of the monitoring system or the portion extending from the confinement boundary to a second isolation valve would have to meet design-basis requirements for accident conditions (i.e., seismic, tipover, and drop loadings). From a practical perspective, external components of the monitoring system (such as tubing, tanks, and pressure gauges) could not readily be designed to prevent confinement rupture during such accident loadings, since they would have to be able to withstand the dynamic crush loading of the cask. However, these accident loadings would not impair the capability of the inner O-ring to maintain the confinement barrier, since it is designed for these loadings and its operation is confirmed through surveillance using the monitoring system.

Therefore, having a monitoring system that is not designed to confinement barrier standards does not pose any significant risk of radioactive gas release from storage. This practice is justified, since the possibility of a design-basis event occurring at a time between surveilances when the inner seal has randomly failed is extremely remote. Other quality assurance (QA) practices associated with fabrication, examination, testing, and inspection of the monitoring system should be commensurate with components of the confinement system.

The monitoring system should be designed so that its failure can readily be identified during routine surveillance. The NRC staff reviews the monitoring system to assess its ability to fulfill its intended function and to determine whether failure of the monitoring system would degrade the safety systems. Although the monitoring system need not remain functional during a particular accident, monitoring capability must be restored following the accident. Consequently, SAR Section 11 should address the corrective action necessary to resume monitoring.

Examine the specified pressure of the gas in the monitored region to verify that it is higher than both the cask cavity and the atmosphere. Coordinate with the structural and thermal reviewers (Chapters 3 and 4 of this SRP) to verify the pressure in the cask cavity.

Review the applicant's analysis to verify that the total volume of gas in the seal monitoring system is such that normal seal leakage will not cause all of the gas to escape over the lifetime of the cask. In determining the proposed maximum leakage rate, the applicant should consider the volume between the redundant seals of the confinement cask, the minimum pressure to be maintained, and the length of the proposed routine recharge cycle. The applicant should then specify the leakage rate as an acceptance test criterion in SAR Section 9, even though the actual leakage rate of the seals is expected to be significantly lower.

For redundant seal welded closures, ensure that the applicant has provided adequate justification that the seal welds have been sufficiently tested and inspected to ensure that the weld will behave similarly to the adjacent parent material of the cask. Any inert gas should not leak or diffuse through the weld and cask material in excess of the design leak rate.

Verify that any leakage test, monitoring, or surveillance conditions are appropriately specified in SAR Sections 9 and 11, the license, and/or the Certificate of Compliance.

3. Nuclides with Potential for Release

The NRC staff has determined that, as a minimum, the nuclides shown below in Table 7.1 must be analyzed for potential accident release. The indicated fractions account for the fact that some of these nuclides will be trapped in the fuel matrix or exist in a chemical or physical form that is not capable of release to the environment under credible accident conditions. The NRC accepts the following fractions available for release from spent fuel from boiling-water reactors (BWRs) and pressurized-water reactors (PWRs) for the purpose of analysis regarding compliance with 10 CFR Part 72. Other accident scenarios may be considered provided the applicant properly justifies the associated release fractions. In some cases the applicant may have to consider other radioactive nuclides depending upon the specific source term analysis of its spent fuel.

The quantities of these radioactive nuclides are often presented in SAR Section 5, since they are generally determined during the evaluation of gamma and neutron source terms in the shielding analysis. Coordinate with the shielding review (Chapter 5 of this SRP) to verify that the applicant has adequately determined these nuclides.

<table>
<tr><th colspan="2" align="center">Table 7.1</th></tr>
<tr><th align="center">Nuclide</th><th align="center">Fractions Available for Release*, 5, 6</th></tr>
<tr><td align="center">^{3}H</td><td align="center">0.30</td></tr>
<tr><td align="center">^{85}Kr</td><td align="center">0.30</td></tr>
<tr><td align="center">^{129}I</td><td align="center">0.10</td></tr>
<tr><td align="center">^{137}Cs</td><td align="center">2.3×10^{-5}</td></tr>
<tr><td align="center">^{134}Cs</td><td align="center">2.3×10^{-5}</td></tr>
<tr><td align="center">^{90}Sr</td><td align="center">2.3×10^{-5}</td></tr>
<tr><td align="center">^{106}Ru</td><td align="center">1.5×10^{-5}</td></tr>
<tr><td align="center">^{60}Co**</td><td align="center">0.15</td></tr>
</table>

* Except for ^{60}Co, only failed fuel rods contribute significantly to the release. Total fraction of radionuclides available for release must be multiplied by the fraction of fuel rods assumed to have failed.

** Source of ^{60}Co is crud on fuel rods, estimated to be 140 µCi/cm^2 for PWRs and 600 µCi/cm^2 for BWRs. Total ^{60}Co activity is this estimate times the total surface area of all rods in the cask[7].

4. Confinement Analyses

Review the applicant's confinement analysis and the resulting annual dose at the controlled area boundary. In general, the staff evaluates analyses for three specific scenarios, as follows:

a. Normal Conditions

If the confinement boundary is welded, or if the region between the two mechanical seals is monitored, the staff accepts that no discernible undetected leakage is credible. Hence, the dose at the controlled area boundary from atmospheric release is negligible.

b. Leakage of One Seal

Depending on its extent, failure of one redundant mechanical seal should not result in release of radioactive material.If the between-seal volume remains at higher pressure than the interior of the cask, no release of radioactive material should ensue. If the pressure differential between the between-seal volume and the atmosphere or the interior of the cask is equalized, however, radioactive material may escape at a rate associated with the acceptable leakage rate across one seal (see V.2, above), the actual pressure differential, and the gas viscosity. Failure of both redundant seals would result in a greater release. Note that components of the between-seal volume pressurization and pressure monitoring system form part of the outer seal.

The NRC staff has accepted this scenario with the assumption that 3 to 10 percent of the fuel rods have failed. Current practice is to assume 10 percent unless the applicant provides sufficient justification for considering a lesser figure. Coordinate with the structural and thermal reviewers (Chapters 3 and 4 of this SRP) to verify that the applicant has adequately determined the cask cavity pressure applicable for this condition.

In addition to the quantity of nuclides available for release and the pressure of the cask cavity, the dose at the controlled area boundary depends on the following factors:

- seal leakage rate

- distance from the cask to the controlled area boundary

- atmospheric dispersion factor

- an individual's breathing rate (except for Kr, for which the dose should be determined using EPA Guide No. 12[8])

- dose conversion factors

The applicant should specify maximum allowable seal leakage rates as design criteria, as discussed in Chapter 12. The minimum distance between the casks and the controlled area boundary is generally also a design criterion; however, 10 CFR Part 72 requires this distance to be at least 100 meters from the ISFSI.

Because a release resulting from seal failure will occur over a substantial period of time, the staff has accepted, as a bounding condition, the atmospheric dispersion factors of Regulatory Guide 1.145[9] on the basis of F-stability diffusion, a wind speed of 1 m/s, and plume meandering. Also, the staff has accepted either an adult breathing rate of 2.5×10^{-4} m^3/s, as specified in Regulatory Guide 1.109[10], or a worker breathing rate of 3.3×10^{-4} m^3/s, as specified in EPA Guidance Report No. 11[11]. Dose conversion factors for inhalation, whole body dose, and thyroid dose should be equivalent to those indicated in EPA Guidance Report No. 11.

Review the applicant's controlled area boundary dose calculation. Verify that the applicant has determined both the whole body dose and the thyroid dose. A conservative bound is established by assuming that an individual is present at the controlled area boundary for the full year (8760 hours). The estimates of the dose that would be received by this individual have typically been low relative to the regulatory limits. An alternative to this conservative assumption may be acceptable if the applicant provides a convincing justification. The dose that an individual would receive in case of a seal leak is usually very small, and this conservatism has not historically posed any difficulties in meeting the regulatory limits; however, this criterion may be reconsidered if the applicant provides sufficient justification.

c. Accident Conditions and Natural Phenomenon Events

Coordinate with the structural reviewers (Chapter 3 of this SRP) to determine the effect of specific accident conditions and natural phenomenon events on the cask confinement system. A full confinement barrier must remain intact under all design-basis accident and natural phenomenon events. Failure of one of the redundant seals may be acceptable as long as the failure of one seal does not result in loss of the confinement function. Nevertheless, to demonstrate the overall safety of dry spent fuel storage, the staff conservatively assumes a failure of the confinement boundary with 100 percent of the fuel rods failed for calculation of an accident dose to a hypothetical individual located at or beyond the boundary of the controlled area.

The analysis for the above scenario is similar to that for failure of one seal. In this situation, the applicant need not consider the cask cavity pressure. Because the leak is assumed to be instantaneous, the plume meandering factor of Regulatory Guide 1.145 is not typically applied. This is equivalent to using an atmospheric dispersion factor based on Regulatory Guide 1.25. Hence, this dispersion factor is generally found to be 4 times higher than that for the case of a single seal failure.

Review the applicant's calculation for the dose at the controlled area boundary, in relation to the regulatory limits listed in 10 CFR 72.106(b). Verify that the applicant has determined both the whole body dose and the thyroid dose. Note that for an instantaneous release (and instantaneous exposure), the time that an individual remains at the controlled area boundary is not a factor in the dose calculation.

5. Supplemental Information

Ensure that all supportive information or documentation has been provided or is readily available. This includes, but is not limited to, justification of assumptions or analytical procedures, test results, photographs, computer program descriptions, input and output, and applicable pages from referenced documents. Reviewers should request any additional information needed to complete the review.

VI. Evaluation Findings

Review the 10 CFR Part 72 acceptance criteria and provide a summary statement for each. These statements should be similar to the following model:

- Section(s) _____ of the SAR describe(s) confinement structures, systems, and components (SSCs) important to safety in sufficient detail in to permit evaluation of their effectiveness.

- The design of the [cask designation] adequately protects the spent fuel cladding against degradation that might otherwise lead to gross ruptures. Section 4 of the safety evaluation report (SER) discusses the relevant temperature considerations.

- The design of the [cask designation] provides redundant sealing of the confinement system closure joints by _____.

- The confinement system is monitored with a _____ monitoring system as discussed above (if applicable). No instrumentation is required to remain operational under accident conditions.

- The quantity of radioactive nuclides postulated to be released to the environment has been assessed as discussed above. In Section 10 of the SER, the dose from these releases will be added to the direct dose to show that the [cask designation] satisfies the regulatory requirements of 10 CFR 72.104(a) and 10 CFR 72.106(b).

- The cask confinement system has been evaluated [by appropriate tests or by other means acceptable to the Commission] to demonstrate that it will reasonably maintain confinement of radioactive material under normal, off-normal, and credible accident conditions.

- The staff concludes that the design of the confinement system of the [cask designation] is in compliance with 10 CFR Part 72 and that the applicable design and acceptance criteria have been satisfied. The evaluation of the confinement system design provides reasonable assurance that the [cask designation] will allow safe storage of spent fuel. This finding is reached on the basis of a review that considered the regulation itself, appropriate regulatory guides, applicable codes and standards, the applicant's analysis and the staff's confirmatory analysis, and accepted engineering practices.

VII. References

1. American Society of Mechanical Engineers, "ASME Boiler and Pressure Vessel Code," Section III, Subsections NB and NC.

2. American National Standards Institute, Institute for Nuclear Materials Management, "American National Standard for Leakage Tests on Packages for Shipment of Radioactive Materials," ANSI N14.5, 1987.

3. Pacific Northwest Laboratory, "Evaluation of Cover Gas Impurities and Their Effects on the Dry Storage of LWR Spent Fuel," PNL-6365, November 1987.

4. Idaho National Engineering Laboratory, "Classification of Transportation Packaging and Dry Spent Fuel Storage System Components According to Importance to Safety," NUREG/CR-6407, INEL-95/0551, February 1996.

5. U.S. Nuclear Regulatory Commission, "Assumptions Used for Evaluating the Potential Radiological Consequences of a Fuel Handling Accident in the Fuel Handling and Storage Facility for Boiling and Pressurized Water Reactors," Regulatory Guide 1.25 (Safety Guide 25), March 1972.

6. E.L. Wilmot, Sandia National Laboratory, "Transportation Accident Scenarios for Commercial Spent Fuel," SAND80-2124, Albuquerque, NM, February 1981.

7. R.P. Sandoval, et al., Sandia National Laboratories, "Estimate of CRUD Contribution to Shipping Cask Containment Requirements," SAND88-1358, TTC-0811, UC-71, January 1991.

8. U.S. Environmental Protection Agency, "Federal Guidance Report No. 12: External Exposure to Radiouclides in Air, Water, and Soil," EPA 402-R-93-081, September 1993.

9. U.S. Nuclear Regulatory Commission, "Atmospheric Dispersement Models for Potential Accident Consequence Assessments at Nuclear Power Plants," Regulatory Guide 1.145, February 1989.

10. U.S. Nuclear Regulatory Commission, "Calculations of Annual Doses to Man from Routine Releases of Reactor Effluents for the Purpose of Evaluating Compliance with 10 CFR Part 50, Appendix I," Regulatory Guide 1.109, October 1977.

11. U.S. Environmental Protection Agency, Federal Guidance Report No. 11, "Limiting Values of Radionuclide Intake and Air Concentration and Dose Conversion Factors for Inhalation, Submersion, and Ingestion," DE89-011065, 1988.

8.0 OPERATING PROCEDURES

I.   Objective

In this portion of the dry cask storage system (DCSS) review, the NRC seeks to ensure that the applicant's safety analysis report (SAR) presents acceptable operating sequences, guidance, and generic procedures for three key operations:

1. cask loading
2. cask handling and storage operations
3. cask unloading

The operating sequences described in the SAR should provide an effective basis for the development of the more detailed operating and test procedures required by the cask user. The user will then use applicant supplied procedures as guidance when preparing and implementing detailed site-specific procedures, as required by the licensee's quality assurance (QA) and procedure writing programs. The NRC normally inspects selected site-specific procedures.

II.   Areas of Review

This chapter of the DCSS Standard Review Plan (SRP) provides guidance in evaluating the applicant's general operating sequences, and generic procedures related to cask operations (i.e., cask loading, cask handling, storage operations, and cask unloading). A comprehensive evaluation of this generic guidance may also encompass those areas of review, as defined in Section V, "Review Procedures." Within each area, the NRC staff assesses the effectiveness of the applicant's generic guidance on a technical and safety basis for the subsequent development of operating detailed procedures. As required by the regulations, [10 CFR 72.234(f)] these procedures are to be provided to each cask user, for the subsequent preparation and implementation of detailed site-specific procedures by the ISFSI licensee.

The purpose of this review,

- loading operations include the selection and placement of fuel into the cask, cask draining and drying, cask decontamination, inerting the cask and sealing the cask

- ISFSI operations include transferring the cask to the ISFSI site and any maintenance or surveillance activities required to ensure the safe storage of the radioactive materials

- Unloading operations required in response to currently unforseen problems that may be encountered during storage or prior to final disposal, including retrieving the cask and preparations for transfer off site

- to recover from an unforeseen problem during storage or to prepare the fuel for offsite transportation or ultimate disposition.

III.   Regulatory Requirements

1. The applicant must develop operating procedures that adequately protect health and minimize danger to life or property. [10 CFR 72.40(a)(5)[1]]

2. The applicant must establish operational restrictions to meet the regulatory requirements of 10 CFR Part 20[2] and objective limits that are as low as is reasonably achievable (ALARA) for radioactive materials in effluents and direct radiation levels associated with ISFSI operations. [10 CFR 72.104(b) and 10 CFR 72.24(e)]

3. The applicant must describe all equipment and processes used to maintain control of radioactive effluents. [10 CFR 72.24(l)(2)]

4. The general licensee shall conduct activities related to storage of spent fuel in accordance with written procedures. [10 CFR 72.212(b)(9)]

5. Vendors seeking approval of a cask design shall ensure that written procedures and appropriate tests are established before initial use of the casks. In addition, the vendor must provide a copy of these

procedures and tests to each prospective cask user. [10 CFR 72.234(f)]

6. The cask must be compatible with wet or dry spent fuel loading and unloading facilities. [10 CFR 72.236(h)]

7. To the extent practicable, the design of the cask must facilitate decontamination. [10 CFR 72.236(i)]

8. The design of storage systems must allow ready retrieval of spent fuel for further processing or disposal. [10 CFR 72.122(l)]

9. The design of the cask must minimize the quantity of radioactive waste generated. [10 CFR 72.128(a)(5) and 10 CFR 72.24(f)]

10. The design of structures, systems, and components (SSCs) that are important to safety must permit inspection, maintenance, and testing. [10 CFR 72.122(f)]

IV. Acceptance Criteria

This SAR section should present a description and identify the sequence of significant operations and actions that are important to safety (i.e., cask loading, cask handling, storage operations, and cask unloading). This section is intended to provide the technical and safety basis for development of the detailed operating procedures prepared by the cask user. Therefore, a sufficient level of detail is needed for the reviewer to conclude that operating procedures derived from the information provided in this section will adequately protect health and minimize danger to life or property, protect the fuel from significant damage or degradation, and provide for the safe performance of tasks and DCSS operations.

To facilitate this conclusion, this portion of the DCSS review seeks to ensure that the generic procedure descriptions and guidance in the SAR include at least the following information:

1. Major operating procedures apply to the principal activities expected to occur during dry cask storage. The expected scope of activities for the SAR operating procedure descriptions is described in Section II, "Areas of Review" (above), as well as Section 8 of Regulatory Guide 3.61[3]. Operating procedure descriptions should be submitted to address the cask design features and planned operations.

2. Operating procedure descriptions should identify measures to control processes and mitigate potential hazards that may be present during planned normal operations. Section V, "Review Procedures" (below), discusses previously identified processes and potential hazards.

3. Operating procedure descriptions should ensure conformance with the applicable operating controls and limits described in the technical specifications provided in SAR Section 12.

4. Operating procedure descriptions should reflect planning to ensure that operations will fulfill the following acceptance criteria:

   a. Occupational radiation exposures will remain ALARA

   b. Effective measures will be taken to preclude potential unplanned and uncontrolled releases of radioactive materials

   c. Offsite dose rates will be maintained within the limits of 10 CFR Part 20 and 10 CFR 72.104 for normal operations, and 10 CFR 72.106 for accident conditions.

      In addition, the operating procedure descriptions should support and be consistent with the bases used to estimate radiation exposures and total doses. (Refer to Chapter 10 of this SRP).

5. Operating procedure descriptions should include provisions for the following activities:

   a. testing, surveillance, and monitoring of the stored material and casks during storage and loading and unloading operations

   b. maintenance of casks and cask functions during storage

c.   contingency actions triggered by inspections, checks, observations, instrument readings, and so forth. (Some of these may involve off-normal conditions addressed in SAR Section 11.)

6.   As required by 10 CFR 72.122(h)(1), the operating procedure descriptions should facilitate reducing the amount of water vapor and oxidizing material within the confinement cask to an acceptable level to protect the spent fuel cladding against degradation that might otherwise lead to gross ruptures.

V.   Review Procedures

The review procedures described in this section are presented in a format intended to facilitate an independent review. Even though several individual(s) may actually be tasked with preparing the section of the safety evaluation report (SER) related to operating procedures, all review team members should examine the operating procedure descriptions presented in the SAR. If the descriptions included in the SAR are not sufficiently detailed to allow a complete evaluation concerning fulfillment of the acceptance criteria, reviewers should request additional information from the applicant.

The operating procedure sequences are described in Section 8 of the SAR, and the direct dose rate information in SAR Section 5 is used to assess compliance with radiation protection requirements in SAR Section 10. The reviewer should verify that the evaluation of Chapter 8 (operating procedures) is coordinated with the shielding and radiation protection evaluations covered in Chapters 5 and 10 of this SRP.

In addition, the following review procedures are based on the assumption that the ISFSI will be located at a reactor facility licensed under 10 CFR Part 50[4] and that loading and unloading activities will be performed in the facility's spent fuel pool. Review procedures for dry fuel transfers and/or ISFSI operations at sites away from a reactor will be developed at a later date.

Reviewers should be familiar with ANSI/ANS 57.9[5] (particularly Appendix A to that standard), which applies to DCSS operating procedures. Background information is available in NUREG/CR-4775[6], which provides guidance on preparing operating procedures for shipping packages. Although NUREG/CR-4775 specifically addresses 10 CFR Part 71[7], most of the guidance can be adapted for storage casks which are governed by 10 CFR Part 72. Consequently, reviewers should be familiar with this information before initiating the DCSS operating procedures review.

Since many of the detailed procedures may be developed by facilities licensed under 10 CFR Part 50 or 72, further background information on site specific procedure requirements may be found in Regulatory Guide 1.33[8] and its associated standard ANSI N18.7/AND 3.2 [9].

In general, reviewers should perform the following steps in the process of evaluating all of the operating procedure descriptions and generic guidance provided in the SAR:

- Verify that the proposed operating procedure descriptions incorporate and are compatible with the applicable operating limits and controls in SAR Section 12, "Conditions for Cask Use." Coordinate with the review of operating controls and limits, as described in Chapter 12 of this SRP.

- Ensure that the proposed operating procedure descriptions properly consider the prevention of hydrogen gas generation from any cause (including the reaction of zinc primer coating with acidic pool water, radiolysis, or other causes). Prevention of hydrogen generation or adequate purging of hydrogen is essential during loading and unloading operations that involve seal welding, seal cutting, grinding, or other forms of hot work.

- Determine whether the descriptions include appropriate precautions to minimize occupational radiation exposures in accordance with ALARA policy and the limits given in 10 CFR Part 20, as mandated by 10 CFR 72.24(e) and 72.126(a)(5). Provisions may include use of remotely controlled equipment, monitoring, and use of portable shielding.

- Verify that the operating procedure descriptions include a general listing of the major tools and equipment needed to support ISFSI loading, storage, and, unloading operations (including those at the pool facility). The descriptions should also address installation, use, and removal of the cask, fuel, tools, equipment. In addition, the descriptions should describe any specialized tools and equipment in sufficient detail to enable users to understand their use and operation.

(Examples include lifting yokes, transporter equipment, welding and cutting equipment, and vacuum drying equipment.) The use of any such equipment, that is classified as being important to safety, is subject to approval as part of the application review. Such equipment should be identified and described in detail; its performance characteristics should be defined; and the design should be evaluated.

In addition to these generic review procedures, reviewers should evaluate each of the specific areas of operating procedure review as described in the following subsections:

1. Cask Loading

The operating procedure descriptions in the SAR should present the activities sequentially in the anticipated order of performance. Review the generic procedures in SAR Section 8 to ensure that they include appropriate key prerequisite, preparation, and receipt inspection activities to be accomplished before cask loading. Also verify that tests, inspections, verifications, and cleaning procedures required in preparation for cask loading are specified. In addition, where applicable, verify that the procedure descriptions include actions needed to ensure that any fluids, such as shield water and primary coolants, fill their respective cavities according to design specifications.

Fuel Specifications

Review the spent fuel specifications (e.g., burnup, cooling period, source terms, heat generation, cladding damage, etc.) in SAR Sections 2 and 12, and verify that the loading procedure description appropriately addresses these specifications. Depending on the types and specifications of fuel assemblies stored in the reactor spent fuel pool, detailed site-specific procedures may be necessary to ensure that all fuel loaded in the cask meets the fuel specifications for the cask design. These procedures can be evaluated only on a site-specific basis and will be generally be evaluated through inspections rather than during the licensing review. The SAR should indicate, however, that such procedures may be necessary.

ALARA

Verify that the procedure descriptions incorporate ALARA principles and practices. These may include provisions to perform radiological surveys, as well as exposure and contamination control measures, temporary shielding, and suggested caution statements related to actions that could change radiological conditions. Verify that any recommended surveys incorporate the applicable operating controls and limits described in SAR Section 12.

Off-site Release

Where applicable, verify that the SAR describes methods to minimize offsite releases such as decontamination, filtered ventilation, temporary containments (tents), and so forth. The procedure descriptions should also provide for minimizing generation of radioactive waste.

Draining and Drying

Evaluate the descriptions related to methods for use in draining and drying the cask. In particular, determine whether they clearly describe the procedures for removing water vapor and oxidizing material to an acceptable level, and assess whether those procedures are appropriate.

The staff has accepted vacuum drying methods comparable to those recommended in PNL-6365[10]. This report evaluates the effects of oxidizing impurities on the dry storage of light-water reactor (LWR) fuel and recommends limiting the maximum quantity of oxidizing gasses (such as O_2, CO_2[a], and CO) to a total of 1 gram-mole per cask. (This corresponds to a concentration of 0.25 volume % of the total gases for a 7.0-m^3 cask gas volume at a pressure of about 0.15 MPa (1.5 atm) at 300°K.) This 1 gram-mole limit reduces the amount of oxidants below levels where any cladding degradation is expected.

Moisture removal is inherent in the vacuum drying process, and levels at or below those evaluated in PNL-6365 (about 0.43 gram-mole H_2O) are expected if adequate vacuum drying is performed. If methods other than vacuum drying are used, review additional analyses to confirm that cover gas

---

[a] Can be broken down by radiolysis

moisture and impurity levels will not result in unacceptable cladding degradation.

The following examples illustrate the accepted methods for cask draining and vacuum drying, in accordance with the recommendations of PNL-6365:

- The cask should be drained of as much water as practicable and evacuated to less than or equal to 4E-4 MPa (3.0 mm Hg or Torr). After evacuation, adequate moisture removal should be verified by maintaining a constant pressure over a period of about 30 minutes without vacuum pump operation. The cask is then backfilled with an inert gas (e.g., helium) for applicable pressure and leak testing. The cask is then re-evacuated and re-backfilled with inert gas before final closure. Care should be taken to preserve the purity of the cover gas and, after backfilling, cover gas purity should be verified by sampling.

- The procedures should reflect the potential for blockage of the evacuation system as a result of icing during evacuation. Icing can occur from the cooling effects of water vaporization and system depressurization during evacuation. Icing is more likely to occur in the evacuation system lines than in the cask because of decay heat from the fuel. A staged draw down or other means of preventing ice blockage of the cask evacuation path may be used (e.g., measurement of cask pressure not involving the line through which the cask is evacuated).

- A suitable inert cover gas (with quality specification that ensures a known maximum of impurities) should be specified to minimize this source of contaminants.

- The process should provide for repetition of the evacuation and repressurization cycles if the cask interior is opened to an oxidizing atmosphere following the evacuation and repressurization cycles (as may occur in conjunction with remedial welding, seal repairs, etc.).

Reviewers should ensure that the vacuum drying specifications are consistent with the proposed operating controls and limits described in the technical specifications provided in SAR Section 12. In addition, reviewers should assess the need for any additional technical specifications.

Filling and Pressurization

Verify that the procedure recommendations address steps to fill and pressurize the cask with inert gas. Also ascertain that the procedure recommendations include the requirements of SAR Section 12.

Ensure that the SAR specifies the leak rate criteria (e.g., total leakage, leakage per closure, sensitivities of tests, and so forth) and that these criteria are consistent with those presented in SAR Sections 2, 9, and 12. Assess whether the general methods of leak testing (e.g., pressure rise, mass spectrometry) apply to the leak rate being tested. Pay particular attention to the possible use of quick-disconnect fittings for draining and filling operations. Although no credit is usually taken for these devices as part of the confinement boundary, their presence can negate the results of the leak test and guidance regarding their use should be provided. In addition, the guidelines presented in the SAR should note that leak testing should be in accordance with ANSI N14.5[11].

Ensure that the SAR presents applicable pressure testing criteria (e.g., test pressure, hold periods, inspections) and that these criteria are consistent with those presented in SAR Section 9.

Welding and Sealing

For seal-welded confinement cask closures, and to ensure ALARA, verify that the SAR specifies the use of a remotely operated welder to make seal welds of the confinement closures. Also verify that the procedures provide for acceptable non-destructive examination of these welds. In addition to leak testing discussed above, the NRC accepts dye penetrant tests on both the root and cover pass of the seal welds on the confinement cask closures (including inner closure; any closure over vent accesses for draining, evacuating, purging, and backfilling the cask interior; and the outer closure).

Verify that the SAR includes acceptable provisions for correction of weld defects and any additional drying and purging that may be necessary. Weld tests should be specified, and be in compliance with descriptions for those tests in the American Society of Mechanical Engineers (ASME) Boiler and Pressure Vessel Code (B&PV)[12].

Verify that provisions for placing and tightening any closure bolts are consistent with information presented in SAR Sections 2, 3, and 9, which address applicable design criteria, structural evaluation, and the acceptance tests and maintenance program, respectively. The inner seal should be tested using a helium leak test with the interior of the cask pressurized as described above. The outer seal should also be tested using a helium leak test with the between-seal volume pressurized as required by the respective subsection of the ASME B&PV Code, Section III.

2. Cask Handling and Storage Operations

Examine the recommendations associated with procedures necessary to transfer the cask to the storage location . Pay particular attention to ensuring that all accident events applicable to such transfer are bounded by the design events analyzed in SAR Sections 2 and 11. Coordinate with the structural and thermal reviews (Chapters 3 and 4 of this SRP) to ensure that all conditions for lifting and handling methods are bounded by the evaluations in SAR Sections 3 and 4. There may be technical specifications associated with cask transfer operations, such as restricting lift heights and environmental conditions (e.g. high/low temperatures, etc.).

Review the procedure recommendations to verify that they discuss the inspection, surveillance, and maintenance requirements that are applicable during ISFSI storage. Surveillance and monitoring requirements should also be included in SAR Section 12, and maintenance should be included in SAR Section 9. Coordinate with the other reviewers to ensure that the operating procedures include the requirements addressed in other sections of the SAR. Note that if the confinement vessel closure is bolted, the staff generally requires that the successful operation of the seals be demonstrated with an initial leak test and a monitoring system and/or a surveillance program, as discussed in Chapter 7 of this SRP.

3. Cask Unloading

Verify that the SAR adequately describes the necessary unloading procedure recommendations. The unloading procedure descriptions should present the activities sequentially in the anticipated order of performance, including those key prerequisite and preparation tasks that must be accomplished before cask unloading. Where applicable, verify that the procedure guidance ensures that any fluids, such as shield or borated water, fill their respective cavities according to design specifications.

Damaged Fuel

The SAR should include appropriate contingency measures for the presence of damaged or oxidized fuel. Procedures should be designed to maximize worker protection from unanticipated radiation exposures or contaminates due to damaged fuel and should implement ALARA procedures, and to the maximum extent possible, prevent any uncontrolled releases to the environment. The following points outline the relevant safety concerns and an acceptable approach to address sampling and damaged fuel contingencies in cask unloading:

- The procedure descriptions should provide for fuel unloading under normal conditions.

- The unloading process must ensure that the fuel can be safely unloaded with regard to structural, criticality, thermal, and radiation protection considerations. This includes the contingency for safe maintenance of the fuel and cask while any additional measures needed to address suspected damaged fuel are planned and implemented.

- The unloading process should reflect the potential for damaged or oxidized fuel and changing radiological conditions.

- The process should include measures to check for and detect damaged or oxidized fuel conditions (such as atmosphere samples) before opening the cask. (Note that fuel oxidation resulting from exposure to air at temperatures typical for dry cask storage is a known form of fuel degradation. Therefore, the presence of air in a cask designed to maintain an inert atmosphere indicates that the fuel may be degraded. The detection of fission gases is another indicator that the fuel may be degraded.)

The process may establish sample result thresholds, above which degraded fuel is suspected. Other technically sound methods may be used to check for potential air leakage paths. Such methods may include designs that monitor cask internal pressure or seal integrity and alert the licensee to a problem before oxidation could occur. However, this method may not address detection of potential fuel degradation resulting from other mechanisms (such as a cask drop accident).

- If the sample indicates normal conditions, the normal unloading process should be followed.

- If degraded fuel is suspected, the procedure description should stipulate that additional measures, appropriate for the specific conditions, are to be planned, reviewed and approved by the designated approval authority, and implemented to minimize exposures to workers and radiological releases to the environment. These additional measures may include provision of filters, respiratory protection, and other methods to control releases and exposures ALARA.

Cooling Venting & Reflooding

Verify that the SAR describes applicable operational measures to control cask cooling, venting, and reflooding. Also verify that these measures are consistent with the results of the structural and thermal evaluations in SAR Sections 3 and 4, respectively. Those evaluations should quantify the applicable design-basis temperatures, allowable pressures, stresses, and material strengths from which the operating controls can be defined.

Coordinate the review of cask cooling, venting, and reflooding measures with the thermal and structural reviews addressed in Chapters 3 and 4 of this SRP. Cask cooling, venting, and reflooding should not cause gross cladding damage. Operational measures may include, but are not limited to, external cooling of the confinement cask for initial temperature reduction, restricting reflood flow rates to control and limit internal cask pressure from steam formation, and limiting cooldown rates.

Special attention should be devoted to reviews in this area, since analysis of existing designs have predicted fuel temperatures during storage and transfer in excess of $500°F$ for design-basis heat loads. Operational controls may be required to address the following potential effects during a cooldown and reflood evolution:

- Cask pressurization may occur as a result of steam formation as reflood water contacts hot surfaces.

- Excessive cooling rates may cause fuel cladding and fuel rod component damage and release of radioactive material as a result of stress (thermal, internal pressure, etc.) beyond material strengths.

- Excessive cooling rates may induce thermal stress that causes gross deformation of the fuel assembly components and subsequent binding with the basket.

- Cask supply and vent line failures from inadequate design for pressure and temperature could result in radiological exposures and personnel hazards (e.g., steam burns).

Fuel Crud

Verify that the procedure descriptions include contingencies for protection from fuel crud particulate material. Appendix B to ANSI/AND 57.9 provides a short discussion of crud with respect to dry transfer systems. However, experience with wet unloading of boiling-water reactor (BWR) fuel after transportation has involved handling significant amounts of crud. This fine crud, includes ^{60}Co and ^{55}Fe, and will remain suspended in water or air for extended periods. The dry cask reflood process during unloading of BWR fuel has the potential to disperse crud into the fuel transfer pool and the pool area atmosphere, thereby creating airborne exposure and personnel contamination hazards. By contrast, no significant crud dispersal problems have been observed in handling pressurized-water reactor (PWR) fuel, because of differences in the characteristics of crud on this type of fuel.

ALARA

Verify that the procedure descriptions incorporate ALARA principles and practices. These may include provisions to perform radiological surveys, exposure and contamination control measures, temporary shielding, and suggested caution statements related to specific actions that could change radiological conditions. Verify that any recommended surveys incorporate the applicable operating controls and limits described in SAR Section 12.

Other

Where applicable, verify that the SAR describes methods (such as filtered ventilation or temporary containments) to minimize offsite releases. The procedures should also provide for minimizing generation of radioactive waste.

VI. Evaluation Findings

Review the 10 CFR Part 72 acceptance criteria and provide a summary statement for each. These statements should be similar to the following model, as applicable:

- The [cask designation] is compatible with [wet/dry] loading and unloading. General procedure descriptions for these operations are summarized in Section(s) _____ of the applicant's safety analysis report (SAR). Detailed procedures will need to be developed and evaluated on a site-specific basis.

- The [bolted lids/other features] of the cask allow ready retrieval of the spent fuel for further processing or disposal as required.

- The smooth surface [or other feature] of the cask is designed to facilitate decontamination. Only routine decontamination will be necessary after the cask is removed from the spent fuel pool.

- No significant radioactive waste is generated during operations associated with the independent spent fuel storage installation (ISFSI). Contaminated water from the spent fuel pool will be governed by the 10 CFR Part 50 license conditions [if applicable].

- No significant radioactive effluents are produced during storage. Any radioactive effluents generated during the cask loading will be governed by the 10 CFR Part 50 license conditions [if applicable].

- The general operating procedures described in the SAR are adequate to protect health and minimize danger to life and property. Detailed procedures will need to be developed and evaluated on a site-specific basis.

- Section 10 of the safety evaluation report (SER) assesses the operational restrictions to meet the limits of 10 CFR Part 20. Additional site-specific restrictions may also be established by the site licensee.

- The staff concludes that the generic procedures and guidance for the operation of the [cask designation] are in compliance with 10 CFR Part 72 and that the applicable acceptance criteria have been satisfied. The evaluation of the operating procedure descriptions provided in the SAR offers reasonable assurance that the cask will enable safe storage of spent fuel. This finding is based on a review that considered the regulations, appropriate regulatory guides, applicable codes and standards, and accepted practices.

VII. References

1. *U.S. Code of Federal Regulations*, "Licensing Requirements for the Independent Storage of Spent Nuclear Fuel and High-level Radioactive Waste," Part 72, Title 10, "Energy."

2. *U.S. Code of Federal Regulations*, Part 20, "Standards for Protection Against Radiation," Title 10, "Energy."

3. U.S. Nuclear Regulatory Commission, "Standard Format and Content for a Topical Safety Analysis Report for a Spent Fuel Dry Storage Cask," Regulatory Guide 3.61, February 1989.

4. *U.S. Code of Federal Regulations*, Part 50, "Domestic Licensing of Production and Utilization Facilities," Title 10, "Energy."

5.   American National Standards Institute, American Nuclear Society, "Design Criteria for an Independent Spent Fuel Storage Installation (Dry Storage Type)," ANSI/ANS 57.9, 1984.

6.   M.C., Witte, Lawrence Livermore National Laboratory, "Guide for Preparing Operating Procedures for Shipping Packages," UCID-20820, NUREG/CR-4775, July 1988.

7.   *U.S. Code of Federal Regulations*, Part 71, "Packaging and Transportation of Radioactive Material," Title 10, "Energy."

8.   U.S. Nuclear Regulatory Commission, "Quality Assurance Program Requirements (Operation)," Regulatory Guide 1.33, February 1978.

9.   American National Standards Institute, American Nuclear Society, "Administrative Controls and Quality Assurance Requirements for the Operational Phase of Nuclear Power Plants," ANSI N18.7/ANS 3.2, 1976.

10. R.W., Knoll, *et al.*, Pacific Northwest Laboratory, "Evaluation of Cover Gas Impurities and Their Effects on the Dry Storage of LWR Spent Fuel," PNL-6365, DE88 003983, November 1987.

11. American National Standards Institute, Institute for Nuclear Materials Management, "American National Standard for Radioactive Materials—Leakage Tests on Packages for Shipment," ANSI N14.5-1987, January 1987.

12. American Society of Mechanical Engineers, "Boiler and Pressure Vessel Code,"

9.0 ACCEPTANCE TESTS AND MAINTENANCE PROGRAM

I   Review Objective

In this portion of the spent fuel dry cask storage system (DCSS) review, the NRC seeks to ensure that the applicant's safety analysis report (SAR) includes the appropriate acceptance tests and maintenance programs for the system. A clear, specific listing of these commitments will help avoid ambiguities concerning design, fabrication, and operational testing requirements when the NRC staff conducts subsequent inspections.

II. Areas of Review

This chapter of the DCSS Standard Review Plan (SRP) provides guidance for use in evaluating the acceptance tests and maintenance programs outlined in the SAR. The acceptance tests demonstrate that the cask has been fabricated in accordance with the design criteria and that the initial operation of the cask complies with regulatory requirements. The maintenance program describes actions that the licensee needs to implement during the storage period to ensure that the cask continues to perform its intended functions.

As defined in Section V, "Review Procedures," a comprehensive evaluation *may* encompass the following acceptance tests and maintenance programs:

1. acceptance tests
    a. visual and nondestructive examination inspections
    b. structural/pressure tests
    c. leak tests
    d. shielding tests
    e. neutron absorber tests
    f. thermal tests
    g. cask identification
2. maintenance program
    a. inspection
    b. tests
    c. repair, replacement, and maintenance

III. REGULATORY REQUIREMENTS

Regulatory requirements applicable to this portion of the DCSS SRP govern the testing and maintenance of the cask system, resolution of issues concerning adequacy or reliability, and cask identification, as follows:

1. Testing and Maintenance

a. The SAR must describe the applicant's program for preoperational testing and initial operations. [10 CFR 72.24(p)[1]]

b. The cask design must permit maintenance as required. [10 CFR 72.236(g)]

c. Structures, systems, and components (SSCs) important to safety must be designed, fabricated, erected, tested, and maintained to quality standards commensurate with the importance to safety of the function they are intended to perform. [10 CFR 72.122(a), 10 CFR 72.122(f), 10 CFR 72.128(a)(1), and 10 CFR 72.24(c)]

d. The applicant or licensee must establish a test program to ensure that all required testing is performed to meet applicable requirements and acceptance criteria. In addition, at least 30 days before the receipt of spent fuel, the licensee must submit to the NRC a report concerning the pre-operational test acceptance criteria and test results. [10 CFR 72.162 and 10 CFR 72.82(e)]

e. The applicant or licensee must evaluate the cask and its systems important to safety, using appropriate tests or other means acceptable to the Commission, to demonstrate that they will

f. The applicant or licensee must inspect the cask to ascertain that there are no cracks, pinholes, uncontrolled voids, or other defects that could significantly reduce confinement effectiveness. [10 CFR 72.236(j)]

g. The applicant must perform, and make provisions that permit the Commission to perform, tests that the Commission deems necessary or appropriate. [10 CFR 72.232(b)]

h. The general licensee must accurately maintain the record provided by the cask supplier showing any maintenance performed on each cask. This record must include evidence that any maintenance and testing have been conducted under an NRC-approved quality assurance (QA) program. [10 CFR 72.212(b)(8)]

i. The applicant or licensee must assure that the casks are conspicuously and durably marked with a model number, unique identification number, and the empty weight [10 CFR 72.236(k)]

2. Resolution of Issues Concerning Adequacy or Reliability

The SAR must identify all SSCs important to safety for which the applicant cannot demonstrate functional adequacy and reliability through previous acceptable evidence. For this purpose, acceptable evidence may be established in any of the following ways:

- prior use for the intended purpose
- reference to widely accepted engineering principles
- reference to performance data in related applications

In addition, the SAR should include a schedule showing how the applicant or licensee will resolve any associated safety questions before the initial receipt of spent fuel. [10 CFR 72.24(i)]

3. Cask Identification

The applicant or licensee must conspicuously and durably mark the cask with a model number, unique identification number, and empty weight. [10 CFR 72.236(k)]

IV. Acceptance Criteria

In general, the acceptance tests and maintenance programs outlined in the SAR should cite appropriate authoritative codes and standards. The staff has previously accepted the following as the regulatory basis for the design, fabrication, inspection, and testing of DCSS components:

SYSTEM/COMPONENT	ACCEPTABLE REGULATORY BASIS*
Confinement System	American Society of Mechanical Engineers (ASME), "Boiler and Pressure Vessel (B&PV) Code"[2], Section III, Subsection NB or NC "American National Standard for Radioactive Materials—Leakage Tests on Packages for Shipment" (ANSI N14.5-1987)[3]
Confinement Internals (e.g., basket)	ASME B&PV Code, Section III, Subsection NG
Metal Cask Overpack	ASME B&PV Code, Section VIII
Concrete Cask Overpack	American Concrete Institute (ACI) Standards 318[4] and 349[5], as appropriate

SYSTEM/COMPONENT	ACCEPTABLE REGULATORY BASIS*
Other Metal Structures	ASME B&PV Code, Section III, Subsection NF American Institute of Steel Construction (AISC), "Manual of Steel Construction"[6]
* The SAR should clearly identify any exceptions to the listed codes and standards.	

In addition, in applications for a dry storage independent spent fuel storage installation (ISFSI), the SAR should cite the applicable design criteria (ANSI/ANS-57.9-1984[7]) promulgated by the American National Standards Institute, American Nuclear Society, which the NRC endorsed via Regulatory Guide 3.60[8]. The SAR should clearly identify any exceptions to this standard, by stating an acceptable method of design for a dry storage ISFSI.

V. Review Procedures

The review procedures described in this section are presented in a format intended to facilitate a single, independent review. Although one or more individual(s) may be tasked with preparing the corresponding section of the safety evaluation report (SER) related to the proposed acceptance tests and maintenance program, all review team members should examine the related information presented in the SAR. Reviewers should devote special attention to those tests (or the lack of tests) that affect their functional area of review. If the descriptions included in the SAR are not sufficiently detailed to allow a complete evaluation concerning fulfillment of the acceptance criteria, reviewers should request additional information from the applicant.

In general, applicants commit to design, construct, and test the system under review to the codes and standards identified in SAR Section 2. The NRC does not generally review specific test procedures as part of the licensing process; however, the applicant is expected to describe (in the SAR) certain elements of the proposed test programs. The staff may inspect selected portions of test procedures as part of its onsite activities.

The following information provides *representative examples* of test program elements that should be subject to licensing review. If included in the SAR, each of these tests should be reviewed to ensure that the applicant has identified the purpose of the test, explained the proposed test method (including any applicable standard to which the test will be performed), defined the acceptance criteria and bases for the test, and described the actions to be taken if the acceptance criteria are not satisfied.

1. Acceptance Tests

The following guidance is presented on the basis of tests deemed acceptable by the staff in previous SAR reviews. Alternative tests and criteria may be used if the SAR provides appropriate explanation and adequate justification.

a. Visual and Nondestructive Examination Inspections

Verify the applicant's commitment to fabricate and examine cask components in accordance with an accepted design standard such as ASME B&PV Code, Section III or VIII. These sections define the examination requirements mentioned in Section II, "Materials Specifications and Properties"; Section V, "NDE Specifications and Procedures"; and Section IX, "Qualification Standard for Welding and Brazing Procedures, Welders, Brazers, and Welding and Brazing Operators." The following guidance assumes that the ASME Code is applicable to the cask being reviewed.

The nondestructive examination (NDE) of weldments must be well-characterized on drawings, using standard NDE symbols and/or notations (see AWS A2.4[9]). Each fabricator should be required to establish and document a detailed, written weld inspection plan in accordance with an approved QA program that complies with 10 CFR Part 72, Subpart G. The inspection plan should include visual (VT), dye penetrant (PT), magnetic particle (MT), ultrasonic (UT), and radiographic (RT) examinations, as applicable. The inspection plan should identify welds to be examined, the examination sequence, type of examination, and the appropriate acceptance criteria as defined by either the ASME code, or an alternative approach proposed and justified by the applicant. Inspection personnel should be pre-qualified, in accordance with the current revision of SNT-TC-1A[10], as specified by the ASME Code. All weld-related NDE should be performed in accordance with written and approved procedures.

Confinement boundary welds and welds for components associated with redundant sealing, must meet the requirements of ASME Code, Section III, Article NB/NC-5200, "Required Examination of Welds." This section generally requires RT for volumetric examination and either PT or MT for surface examination. The ASME-approved specifications for RT, PT, and MT are detailed in ASME Code, Section V, Articles 2, 6, and 7, respectively.

Acceptance criteria for RT should be in accordance with ASME Code, Section III, Subsection NB/NC, Article NB/NC-5320. Testers should reject unacceptable imperfections (such as a crack, a zone of incomplete fusion or penetration, elongated indications with lengths greater than specified limits, and rounded indications in excess of the limits in ASME Code, Section III, Division 1, Appendix VI). Repaired welds should be reexamined in accordance with the original examination method and associated acceptance criteria.

For confinement welds that cannot be volumetrically examined using RT, the licensee may use 100-percent UT. The ASME-approved UT specifications are detailed in ASME Code, Section V, Article 5. Acceptance criteria should be defined in accordance with NB/NC-5330, "Ultrasonic Acceptance Standards." Cracks, lack of fusion, or incomplete penetration are unacceptable, regardless of length.

The NRC has accepted multiple surface examinations of welds, combined with helium leak tests for inspecting the final redundant seal welded closures.

For confinement internals, the licensee should perform all NDE testing in accordance with ASME Code, Section III, Subsection NG.

Nonconfinement welds (which exclude welds of confinement internals) should meet the requirements of ASME Code, Section III, Subsection NF, or Section VIII, Division 1, as applicable. The required volumetric examination of welds is either RT or UT, as discussed in ASME Code, Section III, NF-5200, and Section VIII, UW-11. The appropriate specifications from ASME Code, Section V, are invoked in Article 2 for RT and in Article 5 for UT. Acceptance standards for RT are detailed in ASME Code, Section III, Subsection NF, NF-5320, "Radiographic Acceptance Standards," and for UT in NF-5330, "Ultrasonic Acceptance Standards." For Section VIII weldments, RT acceptance criteria should be in accordance with ASME Code, Section VIII, Division 1, UW-51, and the repair of unacceptable defects should be in accordance with UW-38. Repaired welds should be reexamined in accordance with the original acceptance criteria.

Nonconfinement welds that cannot be examined using RT should undergo UT in accordance with ASME Code, Section V, Article 5. Acceptance criteria should be in accordance with ASME Code, Section VIII, Division 1, UW-53 and Appendix 12, and the repair of unacceptable defects should be in accordance with UW-38. Repaired welds should be reexamined in accordance with the original examination methods and associated acceptance criteria. If applicable, the SAR should also justify the rationale for not requiring RT examination of these welds.

Nonconfinement welds for cask system components that are designed and fabricated in accordance with ASME Code, Section III, that cannot be examined using RT or UT should undergo PT or MT examination in accordance with ASME Code, Section V, Articles 6 and 7, respectively. Acceptance criteria should be in accordance with Articles NF-5350 and NF-5340, respectively. Repaired welds should be reexamined in accordance with the original acceptance criteria. If applicable, the SAR should also justify the rationale for not requiring volumetric inspection techniques (RT or UT) for these welds.

Finished surfaces of the cask should be visually examined in accordance with the ASME Code Section V, Article 9. For welds examined using VT, the acceptance criteria should be in accordance with ASME Code, Section VIII, Division 1, UW-35 and UW-36, or NF-5360, "Acceptance Standards for Visual Examination of Welds".

The licensee should use PT to detect discontinuities (such as cracks, seams, laps, laminations, and porosity) that open to the surface of nonporous metals. PT should be performed in accordance with

ASME Code, Section V, Article 6. Acceptance criteria for confinement welds should be in accordance with ASME Code, Section III, Subsection NB/NC, Article NB/NC-5350. Repair procedures should be in accordance with NB/NC-4450. Acceptance criteria for Nonconfinement welds should be in accordance with ASME Code, Section VIII, Division 1, Appendix 8, or NF-5350, "Liquid Penetrant Acceptance Standards." Repair procedures should be in accordance with ASME Code, Section VIII or NF-2500, "Examination and Repair of Material," and NF-4450, "Repair of Weld Material Defects."

Fabrication controls and specifications should be in-place and field tested to prevent post-welding operations (such as grinding) from compromising the design requirements (such as wall thickness).

b. Structural/Pressure Tests

Lifting trunnions should be fabricated and tested in accordance with ANSI N14.6[11]. Site-specific details of the spent fuel pool and lifting procedures may enable the cask to be considered a non-critical load, as defined by this standard. Generally, however, the cask is considered a critical load during its handling in the spent fuel pool. Consequently, trunnion testing should be performed at a minimum of 150 percent of the maximum service load, if redundant lifting is employed or 300 percent of the service load if non-redundant lifting applies. These load tests should be performed before filling the cask with spent fuel to ensure that the trunnions and cask are conservatively constructed and provide an adequate margin of safety when filled with spent fuel. Trunnion load testing should also be performed annually for the transfer cask and at least one year before use for the storage cask. Load testing of integral trunnions is not required once the loaded storage cask has been placed on the pad. Restrictions on cask lifting resulting from these tests should be included in Section 12 of both the SAR and the related safety evaluation report (SER) prepared by the NRC staff, and SAR Section 9 should explicitly state the testing values.

The confinement boundary (including that of the redundant seal) should be hydrostatically tested to 125 percent of the design pressure, in accordance with ASME Code, Section III, Article NB-6000. (Article NCA-2142.1, defines the design pressure as it applies to Level A Service Limits. As such, in determining the design pressure, the licensee should consider fission gas release from 1 percent of the fuel rods.) The test pressure should be maintained for a minimum of 10 minutes, after which a visual inspection should be performed to detect any leakage. All accessible welds should be inspected using PT. Any evidence of cracking or permanent deformation should constitute cause for rejection. SAR Section 9 should clearly specify the hydrostatic test pressure.

Some casks contain a neutron shielding material that may off-gas at higher temperatures. Such material is usually contained inside a thin steel shell to prevent loss of mass and provide protection from minor accidents and natural phenomenon events. Rupture disks or relief valves are generally provided to prevent catastrophic failure of this shell. The shell should be tested to 125 percent of the rupture disk burst pressure, which is usually equivalent to 125 percent of the shell design pressure. The SAR should clearly specify the burst pressure for the rupture disk, along with its coincident burst temperature and tolerance on burst pressure.

Some cask designs use ferritic steels that are subject to brittle fracture failures at low temperature. ASME Code, Section II, Part A, contains procedures for testing ferritic steel used in low temperature applications. On the basis of guidance in NUREG/CR-1815[12], Section 5.1.1, the NRC established two methods for identifying suitable materials:

- The nil ductility temperature (NDT) must be determined by either direct measurement (ASTM E-208[13]) or indirect measurement (ASTM E-604[14]), and the minimum operating temperature of the steel must be specified as 50°F [(28°C)] higher than the NDT.

- The NRC staff accepts ASME Charpy testing procedures for verification of the material's minimum absorbed energy. Acceptable energy absorption values and test temperatures of Charpy, V-Notch impact tests are listed in the ASME B&PV code Section II, SA-20, "Specifications for General Requirements for Steel Plates for Pressure Vessels" Table A1.15. Coordinate with the thermal review (Chapter 4 of this SRP) to ensure that the applicant selected the correct temperatures for the tests and that the SAR specifies the method of testing.

c. Leak Tests

The licensee should perform leak tests on all boundaries relevant to confinement. These include the primary confinement boundary, the boundary of the redundant seal, and (if applicable) any additional boundaries used in the pressure monitoring system. For all-welded cask confinements, the NRC staff has, with adequate justification, considered it acceptable for licensees to omit leak testing of the second cask closure weld and the seal welds for the closure plates of the purge and vent valves. For such cases, leak testing must show that the inner closure weld meets the leakage limits.

Leakage criteria in units of std cc/s must be at least as restrictive as those specified in the principal design criteria (in SAR Section 2). The SAR should also indicate the general testing methods (e.g., pressure increase, mass spectrometer) and required sensitivities. If cask closure depends on more than one seal (e.g., lid, vent port, drain port), the leakage criteria should ensure that the total leakage is within the design requirements. Leak testing should generally be conducted in accordance with ANSI N14.5.

d. Shielding Tests

Tests of the effectiveness of both the gamma and neutron shielding may be required if, for example, the cask contains a poured lead shield or a special neutron absorbing material. In such instances, the SAR should describe any scanning or probing with an auxiliary source for the purpose of characterizing the shielding. This shield testing should be done for every cask that uses poured shielding material, in order to demonstrate proper fabrication in accordance with the design drawings. The suggested shield test applies equally to both storage and transfer casks.

In addition to the above tests, the licensee should perform dose rate measurements after the spent fuel is loaded, in order to establish that the stated design criteria have been satisfied.

e. Neutron Absorber Tests

The licensee should test fixed neutron absorbers designed to ensure subcriticality, in order to verify the minimum areal density (or other applicable specification) used in the criticality analysis in SAR Section 6. Either the material manufacture or cask fabricator should perform this verification of minimum areal density on a per lot basis.

f. Thermal Tests

Depending on the details of the cask design and the ability to determine its heat removal capability through thermal analysis, testing may be required to verify cask performance. The applicant should establish acceptance criteria on the basis of the conditions of the test (e.g., test heat loading, ambient conditions). The SAR should discuss the correlation between test performance and actual spent fuel loading conditions, in order to avoid ambiguous or unreviewed analysis after the test data are obtained.

g. Cask Identification

The vendor/licensee must mark the cask with a model number, unique identification number, and empty weight. Generally this information will appear on a data plate, which should be detailed in one of the drawings included in SAR Section 1. In addition, vendor/licensee should mark the exterior of shielding casks or other structures that may hold the confinement cask while it is in storage. This marking should provide a unique, permanent, and visible number to permit identification of the cask stored therein.

2. Maintenance Program

Storage casks are typically designed to require minimal maintenance. The SAR should address the following areas, as applicable:

a. Inspection

Usually, the cask has at least one monitoring system (e.g., pressure, temperature, dosimetry). The SAR should discuss how such systems will be used to provide information regarding possible off-normal events and what surveillance actions may be necessary to ensure that these systems function properly. Detailed procedures will be developed and implemented by the licensee at the site.

The SAR should describe routine periodic visual surface and weld inspections, which should be limited to the readily accessible surfaces (i.e., the exterior surface of the storage cask and all surfaces of empty transfer casks). In addition, the SAR should discuss inspection of lifting and rotating trunnion load-bearing surfaces.

b. Tests

The SAR should describe any periodic tests of cask components or calibration of monitoring instrumentation, as well as periodic tests to verify shielding and thermal capabilities. The SAR should also describe procedures for any applicable periodic testing of neutron poison effectiveness. As an alternative to periodic testing of neutron poison effectiveness, the licensee could perform an environmental qualification of the material.

In addition, the SAR should discuss any routine testing of support systems (e.g., vacuum drying, helium backfill, and leak testing equipment).

c. Repair, Replacement, and Maintenance

The SAR should discuss the repair and replacement of cask components, as may be required during the lifetime of the storage and transfer casks. This discussion should include methods of repair or replacement, testing procedures, and acceptance criteria. The SAR should also describe procedures for routine maintenance (such as lubrication and re-application of corrosion inhibiting materials in the event of scratches) through the expiration of the service life of the equipment. Such information is also often included in SAR Section 11, which describes actions to be taken following an off-normal event or accident condition.

VI. Evaluation Findings

Review the 10 CFR Part 72 acceptance criteria and provide a summary statement for each. These statements should be similar to the following model, as applicable:

- Section(s) _____ of the SAR describe(s) the applicant's proposed program for preoperational testing and initial operations of the [cask designation]. Section(s) _____ discuss the proposed maintenance program.

- Structures, systems, and components (SSCs) important to safety will be designed, fabricated, erected, tested, and maintained to quality standards commensurate with the importance to safety of the function they are intended to perform. Section _____ of the SAR identifies the safety importance of SSCs, and Section(s) _____ present(s) the applicable standards for their design, fabrication, and testing.

- The applicant/licensee will examine and/or test the [cask designation] to ensure that it does not exhibit any defects that could significantly reduce its confinement effectiveness. Section(s) _____ of the SAR describe(s) this inspection and testing.

- The applicant/licensee will mark the cask with a data plate indicating its model number, unique identification number, and empty weight. Drawing _____ in SAR Section _____ illustrates and/or describes this data plate.

- The staff concludes that the acceptance tests and maintenance program for the [cask designation] are in compliance with 10 CFR Part 72 and that the applicable acceptance criteria have been satisfied. The evaluation of the acceptance tests and maintenance program provides reasonable assurance that the cask will allow safe storage of spent fuel throughout its licensed or certified term. This finding is reached on the basis of a review that considered the regulation itself, appropriate regulatory guides, applicable codes and standards, and accepted practices.

VII. References

1.   *U.S. Code of Federal Regulations*, Part 72, " Licensing Requirements for the Independent Storage of Spent Nuclear Fuel and High-Level Radioactive Waste," Title 10, "Energy,"

2.   American Society of Mechanical Engineers, "Boiler and Pressure Vessel Code," Sections II, III, V, VIII, and IX, 1992.

3.   American National Standards Institute, Institute for Nuclear Materials Management, "American National Standard for Radioactive Materials Leakage Tests on Packages for Shipment," ANSI N14.5, January 1987.

4.   American Concrete Institute, "Code Requirements for Structural Plain Concrete," ACI-318.

5.   American Concrete Institute, "Code Requirements for Nuclear Safety Related Concrete Structures," ACI-349.

6.   American Institute of Steel Construction, "Manual of Steel Construction."

7.   American National Standards Institute, American Nuclear Society, "Design Criteria for an Independent Spent Fuel Storage Installation (Dry Storage Type)," ANSI/ANS-57.9, 1984.

8.   U.S. Nuclear Regulatory Commission, "Design of an Independent Spent Fuel Storage Installation (Dry Storage)," Regulatory Guide 3.60, March 1987.

9.   American Welding Society, "Standard Symbols for Welding, Brazing, and Nondestructive Examination," AWS A2.4, 1993.

10.   American Society for Nondestructive Testing, "Personnel Qualification and Certification in Nondestructive Testing ," Recommended Practice No. SNT-TC-1A, December 1992.

11.   American National Standards Institute, Institute for Nuclear Materials Management, "American National Standard for Radioactive Materials—Special Lifting Devices for Shipping Containers Weighing 10,000 Pounds (4500 Kilograms) or More," ANSI N14.6, September 1986.

12.   W.R. Holman and R.T. Langland, Lawrence Livermore Laboratory, "Recommendations for Protecting Against Failure by Brittle Fracture in Ferritic Steel Shipping Containers Up to Four Inches Thick," NUREG\CR-1815, August 1981.

13.   American Society for Testing and Materials, "Method of Conducting Drop Weight Test to Determine Nil-Ductility Transition Temperature for Ferritc Steel ," ASTM E-208.

14.   American Society for Testing and Materials, "Dynamic Tear Testing of Metallic Materials," ASTM E-604.

10.0 RADIATION PROTECTION

I.   Review Objective

In this portion of the dry cask storage system (DCSS) review, the NRC evaluates the radiation protection capabilities of the proposed cask system. In particular, the NRC staff considers the following aspects:

- Do the proposed DCSS radiation protection features meet the NRC's design criteria for direct radiation?

- Has the applicant proposed engineering features and operating procedures for the DCSS that will ensure the worker's exposures remain as low as is reasonably achievable (ALARA)?

- Will the radiation doses to the general public meet regulatory standards during both normal operation and accident situations?

In ISFSI operation, the major mode of radiation exposure associated with spent fuel storage cask handling results from direct radiation. Because of the cask design requirements, radionuclides are not expected to be released from the cask during either normal operations or design-basis accidents (DBAs).

II.   Areas of Review

This chapter of the DCSS Standard Review Plan (SRP) provides guidance for use in evaluating the radiation protection capabilities of the proposed cask system. As defined in Section V, "Review Procedures," a comprehensive radiation protection evaluation *may* encompass the following areas of review:

1. radiation protection design criteria and features
2. occupational exposures
3. public exposures
    a.   normal conditions
    b.   accident conditions and natural phenomenon events
4. ALARA

III. Regulatory Requirements

1. Criteria for radioactive material released due to effluents and direct radiation from an ISFSI or MRS are contained 10 CFR 72.104[1].

2. Criteria for Occupational Exposures are contained in 10 CFR 20.1201, 10 CFR 20.1207, 10 CFR 20.1208, and 10 CFR 20.1301

3. Criteria for public exposures under normal and accident conditions are contained within. [10 CFR 72.104 and 10 CFR 72.106]

4. Criteria for ALARA are contained within 10 CFR 20.1101, 10 CFR 72.24(e), 10 CFR 72.104(b), and 10 CFR 72.126(a)]

IV. Acceptance Criteria

In general, the DCSS radiation protection evaluation seeks to ensure that the proposed design fulfills the following acceptance criteria:

1. Design Criteria

Limitations on dose rates associated with direct radiation from the cask are established on the basis of the shielding and confinement evaluations in order to satisfy the regulatory requirements for public dose limits. As stated in 10 CFR Part 72.104, during normal operations and anticipated occurrences, the annual dose equivalent to a real individual located beyond the controlled area, must not exceed the limits discussed below.

2. **Occupational Exposures**

    a. dose limits for adults: 5 rem/yr (total effective dose equivalent)

    b. dose limits for minors: 0.5 rem/yr

    c. dose to an embryo or fetus
       (declared pregnant woman):   0.5 rem during entire pregnancy

3. **Public Exposures**

    a. **Normal Conditions:**

whole body:	25 mrem/yr
thyroid:	75 mrem/yr
other organ:	25 mrem/yr

These doses include the cumulative effects of other nuclear fuel cycle facilities that may be at the same location as the storage system (i.e., the nuclear power plant) and apply to the limiting real individual of the general public residing at a permanent location nearest the facility.

    b. **Accident Conditions and Natural Phenomenon Events**

        5 rem to the whole body or any organ of any individual located at or beyond the nearest boundary of the controlled area.

4. **ALARA**

As a minimum, the proposed ALARA policy must fulfill the following criteria:

    a. To the extent practicable, the applicant should employ procedures and engineering controls that are founded upon sound radiation protection principles.

    b. Any design change should account for radiation protection, technological, and economical considerations.

    c. The applicant should have a written policy statement reflecting management commitment to maintain occupational and public exposures to radiation and radioactive material ALARA.

V. Review Procedures

1. **Radiation Protection Design Criteria and Features**

    a. **Design Criteria**

Review the principal design criteria presented in Chapter 1 of the applicant's safety analysis report (SAR), as well as any additional detail regarding radiation protection provided in the Shielding and Confinement Evaluation sections of the SAR. Additional criteria that should be presented in SAR Section 10 (if not previously discussed) include (but are not limited to) the following:

    (1) The cask system design must satisfy the ALARA and other occupational exposure requirements of 10 CFR Part 20[2].

    (2) The sum of the doses from direct radiation and from release of nuclides to the atmosphere must satisfy the requirements of 10 CFR 72.104(a) and 72.106(b). Because of the stringent design requirements for spent fuel cask systems, the release of nuclides into the atmosphere is expected to be insignificant under both normal and accident conditions. Direct radiation is the major mode of exposure.

b. Design Features

Review the general description and functional features of the cask presented in the General Description, as well as any additional information provided in the Shielding and Confinement Evaluation sections of the SAR. In general, the applicant's approach to the relevant design criteria are discussed in these earlier sections of the SAR. They may also be summarily noted in the Radiation Protection section of the safety evaluation report (SER) prepared by the NRC staff.

2. Occupational Exposures

Review the operating procedures in SER Section 8 and direct radiation dose calculations in SER Section 5. The applicant should use these data in SER Section 10 to estimate the doses received by occupational personnel during cask loading and transportation to the ISFSI. The applicant should also identify any significant differences from these doses that may occur during cask retrieval and unloading. In addition, the applicant should present similar dose estimates for periodic or routine maintenance, as well as surveillance activities. These estimates may require additional assumptions concerning adjacent casks for a typical storage configuration.

Also in SAR Section 10, the applicant should present the rationale used to justify the bases for the various exposure times, personnel locations relative to the casks (including hot spots), number of personnel required, and appropriate gamma and neutron dose rates. Verify that the calculated doses are consistent with these estimates. Keep in mind that the actual operations will be performed under an active dose monitoring program that further ensures compliance with the requirements of 10 CFR Part 20. NRC Regulatory Guide (RG) 8.34[3], which was developed to implement revisions to 10 CFR Part 20, can be used to determine the acceptability of the applicant's occupational exposure evaluation and monitoring recommendations.

3. Public Exposures

An SAR for an application seeking approval of a DCSS under 10 CFR Part 72, Subpart L, should include an analysis of potential public exposures that will facilitate a future site-specific suitability analysis required by a licensee prior to DCSS use. One approach is for the applicant to include a dose rate versus distance curve for an assumed array of casks. This curve would assist the reviewer in the determination of the cumulative exposure effects. As an alternative, the analyses documented in the SAR may presume that the public exposure occurs at a distance of 100 meters from the closest stored fuel, with the most severe concentration of casks, and a distance of at least 100 meters between the transfer path and the closest point of public access. 10 CFR 72.106(b) specifies 100 meters as the minimum distance to the closest boundary of the controlled area. These assumptions should be conservative relative to most actual site conditions.

a. Normal Conditions

Review the information in SAR Section 5 regarding the direct dose rate at the controlled area boundary. For applications requesting approval of a cask system under 10 CFR Part 72, Subpart L, the dose for the public should be determined at a distance of 100 meters from the closest boundary of the controlled area, as specified in 10 CFR 72.106(b). However, the applicant may use a longer distance, provided that the longer distance is made a condition of use.

The sum of doses, including an additional margin to account for doses received from other fuel cycle (reactor) operations, must satisfy the requirements of 10 CFR 72.104(a). As discussed in Chapter 5 of this SRP, the direct dose at the controlled area boundary depends on many site-specific conditions, which the SAR may treat in a general manner. Verify that the SAR includes a requirement for site-specific dose analysis and monitoring by the ISFSI licensee, or that the applicant has presented sufficient bounding analyses. (The latter approach will generally require extensive calculations.)

b. Accident Conditions and Natural Phenomenon Events

Review the direct dose rate associated with accident conditions at the boundary of the controlled area, as discussed in Chapter 5. Also review the dose rate resulting from accidental release of radionuclides, as presented in Chapter 7. The accident-related radionuclide release dose should account for both air and liquid pathways as appropriate. In addition, verify that the applicant has evaluated the source terms for both spent fuel fission product and cask surface contamination. The sum of these must satisfy the requirements of 10 CFR 72.106(b). For purposes of demonstrating compliance with 10 CFR 72.106(b), and evaluation against the Environmental Protection Agency Protective Action Guides[4], the skin, extremities, and the lens of the eye may be considered separately from other organs.

As noted in Chapter 5 of this SRP, the time-integrated direct dose at the boundary of the controlled area may be small (compared with that of a hypothetical instantaneous release of all available fission product gases). Consequently, the applicant should estimate the doses at a distance of 100 meters from the storage location to the nearest boundary of the controlled area, unless the SAR specifies a greater distance that is also made a condition of use for the proposed DCSS. Alternatively, applicants may depict dose estimation using a curve showing dose versus distance from an assumed array of casks

4. ALARA

Review the applicant's stated commitment to ALARA policy, and determine whether this commitment influenced the proposed cask design features and operating procedures.

To determine if the applicant's ALARA policy is acceptable, review the evidence that the design methods, approaches, and interactions are in accordance with the ALARA provision in Regulatory Guides 8.8[5] and 8.10[6].

VI. Evaluation Findings

Review the acceptance criteria in Chapter 8.IV of this SRP and provide a summary statement for each. These statements should be similar to the following model:

- The [cask designation] provides radiation shielding and confinement features that are sufficient to meet the requirements of 10 CFR 72.104 and 72.106.

- Occupational radiation exposures satisfy the limits of 10 CFR Part 20 and meet the objective of maintaining exposures ALARA.

- The staff concludes that the design of the radiation protection system of the [cask designation] is in compliance with 10 CFR Part 72 and that the applicable design and acceptance criteria have been satisfied. The evaluation of the radiation protection system design provides reasonable assurance that the [cask designation] will allow safe storage of spent fuel. This finding is reached on the basis of a review that considered the regulation itself, appropriate regulatory guides, applicable codes and standards, and accepted engineering practices.

VII. References

1. *U.S. Code of Federal Regulations*, "Licensing Requirements for the Independent Storage of Spent Nuclear Fuel and High-level Radioactive Waste," Part 72, Title 10, "Energy."

2. *U.S. Code of Federal Regulations*, Part 20, "Standards for Protection Against Radiation," Title 10, "Energy."

3. U.S. Nuclear Regulatory Commission, "Monitoring Criteria and Methods to Calculate Occupational Radiation Doses," Regulatory Guide 8.34, July 1992.

4. Environmental Protection Agency, "Manual of Protective Action Guides and Protective Actions for Nuclear Incidents", EPA 410-R-92-001, May 1992.

5. U.S. Nuclear Regulatory Commission, "Information Relevant to Ensuring that Occupational Radiation Exposures at Nuclear Power Stations Will Be As Low As Reasonably Achievable," Regulatory Guide 8.8, Rev. 3, June 1978.

6. U.S. Nuclear Regulatory Commission, "Operating Philosophy for Maintaining Occupational Radiation Exposures As Low As is Reasonably Achievable," Regulatory Guide 8.10, Revision 1-R, May 1977.

11.0 ACCIDENT ANALYSES

I.   Review Objective

In this portion of the dry cask storage system (DCSS) review, the NRC evaluates the applicant's identification and analysis of hazards, as well as the summary analysis of system responses to both off-normal and accident or design-basis events. This review ensures that the applicant has conducted thorough accident analyses, as reflected by the following factors:

1.   identified all credible accidents
2.   provided complete information in the safety analysis report (SAR)
3.   analyzed the safety performance of the cask system in each review area
4.   fulfilled all applicable regulatory requirements

II.   Areas of Review

This portion of the DCSS review evaluates the applicant's identification and analysis of hazards, with particular emphasis on the safety performance of the cask system under off-normal events and conditions and accident or design-basis events. Consequently, this chapter of the DCSS Standard Review Plan (SRP) provides guidance for use in reviewing the applicant's identification and analysis of hazards, as well as the summary analysis of system responses. A comprehensive accident analysis evaluation *may* encompass the following areas of review:

1.   cause of the event
2.   detection of the event
3.   summary of event consequences and regulatory compliance
4.   corrective course of action

III.   Regulatory Requirements

1.   Structures, systems, and components (SSC) important to safety must be designed to withstand credible accidents and natural phenomena without impairing their ability to perform safety functions. [10 CFR 72.24(d)(2); 10 CFR 72.122(b)(2), (3), (d), and (g)]

2.   During normal operations and anticipated occurrences, the annual dose equivalent to any real individual who is located beyond the controlled area must not exceed 25 mrem to the whole body, 75 mrem to the thyroid and 25 mrem to any other organ as a result of exposure to the sources listed in the regulations. [10 CFR 72.104(a); 10 CFR 72.236(d); and 10 CFR 72.24(d)]

3.   Dose Limits for Design-Basis Accidents require that any individual located on or beyond the nearest boundary of the controlled area shall not receive a dose greater than 5 rem to the whole body or any organ from any design basis accident. [10 CFR 72.106(b); 10 CFR 72.24(m); and 10 CFR 72.24(d)(2)]

4.   The spent fuel must be maintained in a subcritical condition under credible conditions [10 CFR 72.236(c) and 10 CFR 72.124(a)]

5.   The cask and its systems important to safety must be evaluated, using appropriate tests or by other means acceptable to the Commission, to demonstrate that they will reasonably maintain confinement of radioactive material under credible accident conditions [10 CFR 72.236(l)]

6.   Storage systems must allow ready retrieval of spent fuel for further processing or disposal. [10 CFR 72.122(l)]

7.   Instrumentation and control systems must be provided to monitor systems that are important to safety over anticipated ranges for normal operation and off-normal operation. Those instruments and control systems that must remain operational under accident conditions must be identified in the Safety Analysis Report [10 CFR 72.122(i)]

8. Where Instrumentation and control systems are not appropriate. Storage confinement systems must have the capability for continuous monitoring in a manner such that the licensee will be able to determine when corrective action needs to be taken to maintain safe storage conditions. [72.122(h)(4)]

IV. Acceptance Criteria

Accidents and events associated with natural phenomena may share common regulatory and design limits. Consequently, the following sections sometimes refer to these scenarios collectively as *accident conditions*.

By contrast, anticipated occurrences (off-normal conditions) are distinguished, in part, from accidents or natural phenomena by the appropriate regulatory guidance and design criteria. For example, the radiation dose from an off-normal event must not exceed the limits specified in 10 CFR Part 20 and 10 CFR 72.104(a), whereas the radiation dose from an accident or natural phenomenon must not exceed the specifications of 10 CFR 72.106(b). Accident conditions may also have different allowable structural criteria.

In general, this portion of the DCSS review seeks to ensure that the given design and the applicant's hazard identification and analyses of related system responses fulfill the following acceptance criteria:

1. Dose Limits for Off-Normal Events

During normal operations and anticipated occurrences, the requirements specified in 10 CFR Part 20 must be met. In addition the annual dose equivalent to any individual located beyond the controlled area must not exceed 25 mrem to the whole body, 75 mrem to the thyroid, and 25 mrem to any other organ as a result of exposure to the following sources:

   a. planned discharges to the general environment of radioactive materials (with the exception of radon and its decay products)

   b. direct radiation from operations of the independent spent fuel storage installation (ISFSI)

   c. any other cumulative radiation from uranium fuel cycle operations (i.e., nuclear power plant) in the affected area

2. Dose Limit for Design-Basis Accidents

Any individual located at or beyond the nearest controlled area boundary must not receive a dose greater than 5 rem to the whole body or any organ from any design-basis accident.

3. Criticality

The spent fuel must be maintained in a subcritical condition under credible conditions (i.e., k_{eff} equal to or less than 0.95). At least two unlikely, independent, and concurrent or sequential changes must be postulated to occur in the conditions essential to nuclear criticality safety before a nuclear criticality accident is possible (double contingency).

4. Confinement

The cask and its systems important to safety must be evaluated, using appropriate tests or by other means acceptable to the Commission, to demonstrate that they will reasonably maintain confinement of radioactive material under credible accident conditions.

5. Retrievability

Retrievability is the capability to return the stored radioactive material to a safe condition without endangering public health and safety. This generally means ensuring that any potential release of radioactive materials to the environment or radiation exposures is not in excess of the limits in 10 CFR 20[2] or 10 CFR 72.122(h)(5). ISFSI and MRS storage systems must be designed to allow ready retrieval of the stored spent fuel or high level waste (MRS only) for compliance with 10 CFR 72.122(l).

6. Instrumentation

The SAR must identify all instruments and control systems that must remain operational under accident conditions.

V.   Review Procedures

More detailed review procedures are under staff consideration and will be provided in a future revision to this SRP.

The review procedures presented here describe general procedures for reviewing a DCSS submittal.

Review the off-normal conditions, accidents, and natural phenomena events identified in SAR Section 2. For each type of event, this discussion should include the applicant's evaluation of the following areas, as applicable.

1.   Cause of the Event

In some cases, an event may be analyzed for regulatory purposes even though no credible cause can be identified. Such events should be clearly identified as "non-mechanistic."

2.   Detection of the Event

The licensee may detect an event through visual surveillance or monitoring instrumentation and alarms. The method of detection will be intuitively obvious for some events, whereas other events (e.g., fuel rod rupture) may remain undetected, at least for a significant period of time.

DCSS monitoring equipment (such as a pressure monitoring system) would be classified as important to safety in accordance with NUREG/CR-6407[3], "Classification of Transportation Packaging and Dry Spent Fuel Storage System Components According to Importance to Safety." Refer to Chapter 7 of this SRP.

Plant monitoring equipment (such as the seismic monitoring system) used by the ISFSI would be classified in accordance with the regulations of 10 CFR 50.55a and guidance provided in NUREG-0800[4], "SRP for the Review of SAR for Nuclear Power Plants," Section 7.5, "Information Systems Important to Safety."

Note that if the licensed power plant is decommissioned, any plant monitoring equipment that is relied on by the ISFSI must be added to the ISFSI monitoring equipment.

3.   Summary of Event Consequences and Regulatory Compliance

The applicant should address event consequences in each functional area corresponding to earlier sections of the SAR (i.e., structural, thermal, shielding, criticality, confinement, and radiation protection). This discussion should refer back to each SAR section in which the individual consequences are evaluated in detail. In addition, the applicant should show that the consequences comply with the applicable regulatory criteria. For anticipated occurrences, the applicant should demonstrate compliance with Part 20 as well as Part 72

Because of the stringent design requirements for dry storage casks, it is expected that significant releases of radioactive material will not occur under normal, off-normal, and design-basis accident conditions; the major source of exposure is expected to be direct radiation. However, to demonstrate the overall safety of the cask storage system, and to illustrate compliance with regulatory limits, the applicant's analyses should presume that a non-mechanistic event will result in a release of radioactive material. This approach is normally applied in the environmental assessment, to illustrate compliance with regulatory dose requirements, and to aid the licensee in establishing an appropriate controlled area boundary.

4.   Corrective Course of Action

The applicant should identify what action(s), if any, would be necessary to recover from the event. If various courses of action are possible, the applicant should present a discussion concerning the selection of the most appropriate action. Because the fuel must be readily retrievable, returning the cask to the fuel handling building and reloading the spent fuel into a new cask is a viable option.

The primary emphasis in this portion of the DCSS review is to assess whether the applicant has provided complete information regarding the accident analyses. Therefore, the individual reviewers should ensure that the applicant has identified and analyzed credible situations, addressed their impact on each review area, and satisfied the applicable regulatory criteria.

VI. Evaluation Findings

Review the 10 CFR Part 72 acceptance criteria and provide a summary statement for each. These statements should be similar to the following model:

- Structures, systems, and components of the [cask designation] are adequate to prevent accidents and to mitigate the consequences of accidents and natural phenomena events that do occur.

- The spacing of casks, discussed in Section _____ of the safety evaluation report (SER) and included as an operating limit in Section 12 of the SAR will ensure accessibility of the equipment and services required for emergency response.

- Table ___ of the SER lists the Technical Specifications for the [cask system designation]. These Technical Specifications are further discussed in Section _____ of the SER.

- The applicant has evaluated the [cask designation] to demonstrate that it will reasonably maintain confinement of radioactive material under credible accident conditions.

- An accident or natural phenomena event will not preclude the ready retrieval of spent fuel for further processing or disposal.

- The spent fuel will be maintained in a subcritical condition under accident conditions.

- Neither off-normal nor accident conditions will result in a dose, to an individual outside the controlled area, that exceeds the limits of 10 CFR 72.104(a) or 72.106(b), respectively.

- No instruments or control systems are required to remain operational under accident conditions [as applicable].

The staff concludes that the accident design criteria for the [cask designation] are in compliance with 10 CFR Part 72 and the accident design and acceptance criteria have been satisfied. The applicant's accident evaluation of the cask adequately demonstrates that it will provide for safe storage of spent fuel during credible accident situations. This finding is reached on the basis of a review that considered independent confirmatory calculations, the regulation itself, appropriate regulatory guides, applicable codes and standards, and accepted engineering practices.

VII. References

1. *U.S. Code of Federal Regulations*, "Licensing Requirements for the Independent Storage of Spent Nuclear Fuel and High-level Radioactive Waste," Part 72, Title 10, "Energy."

2. *U.S. Code of Federal Regulations*, Part 20, "Standards for Protection Against Radiation," Title 10, "Energy."

3. NUREG/CR 6407, "Classification of Transportation Packaging and Dry Spent Fuel Storage System Components According to Importance to Safety," February 1996.

4. U.S. Nuclear Regulatory Commission, NUREG-0800; "Standard Review Plan for the Review of Safety Analysis Reports for Nuclear Power Plants,"

12.0 CONDITIONS FOR CASK USE —
OPERATING CONTROLS AND LIMITS OR TECHNICAL
SPECIFICATIONS

I. Review Objective

In this portion of the dry cask storage system (DCSS) review, the NRC evaluates the operating controls and limits or the technical specifications (including their bases and justification) that the applicant has established as license conditions (for site-specific applications) or conditions of use (for applications requesting DCSS system approval under 10 CFR Part 72[1]). The NRC also determines whether the applicant has fully evaluated the proposed operating controls and limits, or technical specifications, and whether the safety evaluation report (SER) prepared by the NRC staff incorporates any additional operating controls and limits that the staff deems necessary.

For simplicity in defining the acceptance criteria and review procedures, the term *technical specifications* may be considered synonymous with *operating controls and limits*. The technical specifications define the conditions that are deemed necessary for safe DCSS system use. Specifically, they define operating limits and controls, monitoring instruments and control settings, surveillance requirements, design features, and administrative controls that ensure safe operation of the DCSS. As such, these technical specifications are included in a DCSS certificate of compliance or site-specific license, as applicable. Each specification should be clearly documented and justified in the technical review sections of the safety analysis report (SAR) and the associated SER as necessary for safe DCSS operation.

II. Areas of Review

This chapter of the DCSS Standard Review Plan (SRP) provides guidance for use in evaluating the technical specifications that the applicant deems necessary for safe use of the proposed DCSS system. As defined in Section V, "Review Procedures," a comprehensive review of the proposed technical specifications would assess the applicant's compliance with the regulatory requirements for technical specifications, as defined in 10 CFR Part 72.44(c). These requirements represent the following five areas of review:

1. functional/operating limits, monitoring instruments, and limiting control settings
2. limiting conditions
3. surveillance requirements
4. design features
5. administrative controls

III. Regulatory Requirements

1. General Requirement for Technical Specifications

The applicant shall propose technical specifications (complete with acceptable bases and adequate justification). These specifications must include the following five areas [10 CFR 72.44(c), 10 CFR 72.24(g), and 10 CFR 72.26]:

    a. functional/operating limits, monitoring instruments, and limiting controls
    b. limiting conditions
    c. surveillance requirements
    d. design features
    e. administrative controls

Subpart E, "Siting Evaluation Factors," and Subpart F, "General Design Criteria," to 10 CFR Part 72, provide the bases for the cask system design and, hence, are applicable as bases for appropriate technical specifications.

2. Specific Requirements for Technical Specifications — Storage Cask Approval

As a condition of approval, the design, fabrication, testing, and maintenance of a spent fuel DCSS must comply with the requirements of 10 CFR 72.236. [10 CFR 72.234(a)]

NUREG-1536

The applicant must provide specifications for the spent fuel to be stored in the DCSS. At a minimum, these specifications should include, but not be limited to the following details [10 CFR 72.236(a)]:

    a. type of spent fuel (i.e., BWR, PWR, or both)
    b. maximum allowable enrichment of the fuel prior to any irradiation
    c. burn-up (i.e., megawatt-days/MTU)
    d. minimum acceptable cooling time of the spent fuel prior to storage in the DCSS (minimum 1 year)
    e. maximum heat that the DCSS system is designed to dissipate
    f. maximum spent fuel loading limit
    g. weights and dimensions
    h. condition of the spent fuel (i.e., intact assembly or consolidated fuel rods)
    i. inerting atmosphere requirements

The applicant must provide design bases and design criteria for structures, systems, and components (SSCs) important to safety. [10 CFR 72.236(b)]

The applicant must design and fabricate the DCSS so that the spent fuel will be maintained in a subcritical condition under credible conditions. [10 CFR 72.236(c)]

The applicant must provide radiation shielding and confinement features that are sufficient to meet the requirements in 10 CFR 72.104 and 72.106 regarding radioactive material in effluents, direct radiation, and area control. [10 CFR 72.236(d) and 10 CFR Part 20[2]]

The applicant must design the DCSS to meet the following criteria:

- Provide redundant sealing of confinement systems. [10 CFR 72.236(e)]

- Provide adequate heat removal capacity without active cooling systems. [10 CFR 72.236(f)]

- Safely store the spent fuel for a minimum of 20 years and permit maintenance as required. [10 CFR 72.236(g)]

- Facilitate decontamination to the extent practicable. [10 CFR 72.236(i)]

The DCSS must be compatible with wet or dry spent fuel loading and unloading facilities. [10 CFR 72.236(h)]

The applicant must inspect the DCSS to ascertain that there are no cracks, pinholes, uncontrolled voids, or other defects that could significantly reduce its confinement effectiveness. [10 CFR 72.236(j)]

The applicant must evaluate the DCSS, and its systems important to safety, using appropriate tests or other means acceptable to the Commission, to demonstrate that they will reasonably maintain confinement of radioactive material under normal, off-normal, and credible accident conditions. [10 CFR 72.236(l)]

IV.   Acceptance Criteria

In this portion of the DCSS review, the NRC evaluates the technical specifications that the applicant deems necessary and sufficient for safe use of the proposed DCSS. This evaluation is based on information that the applicant presents in SAR Section 12[*], as well as accepted practices and applicant commitments discussed in other sections of the SAR or in correspondence subsequent to submission of the application.

The staff generally expects that appropriate section of the SAR will include commitments regarding DCSS design and use. Where this expectation is not met, the staff may request that the applicant revise the SAR to include certain commitments. Such commitments may also be documented in other pertinent

---

[*] An applicant may submit the proposed operating controls, limits and technical specifications as a separate document, provided that the document is submitted at the same time as the application.

applicant correspondence, or in site-specific license conditions as required by applicable regulations. Nonetheless, it is helpful to ensure that SAR Section 12 includes *all* commitments applicable to technical specifications (regardless of where else they may be addressed), and that the SER prepared by the NRC staff acknowledges each of these commitments. Moreover, it should be noted that the fact that SAR Section 12 does not explicitly address all commitments does not negate the requirement of the DCSS owner/operator to meet and comply with such commitments.

Because of the breadth and scope of the conditions for DCSS use, it is not possible to define each instance where a technical specification is necessary. The applicant is responsible for submitting a complete application that proposes all conditions that are deemed necessary for safe DCSS use. For these reasons, it is important that NRC staff reviewers conduct a detailed, thorough, and independent evaluation of each technical section and its associated technical specifications. In particular, the pertinent SAR sections must identify and support the technical specifications deemed necessary to maintain subcriticality, confinement barrier integrity, shielding and radiological protection, heat removal capability, and structural integrity under normal and accident operations.

V. Review Procedures

Evaluate each section of the SAR with the goal of establishing the technical specifications. Each reviewer of the SAR should note all instances in which the SAR either makes an assumption or imposes a condition that should be identified as a technical specification. Reviewers should also note any instances in which the SAR requests exceptions or exemptions from regulatory requirements, or other conditions that the reviewer identifies as an operational limit or condition. Reviewers assigned to this portion of the review should also ensure that such limits and exemptions are clearly identified and documented in SAR Section 12, and acknowledged in SER Section 12.

Review or be familiar with the technical specifications of similar cask designs previously approved by the NRC staff. For example, the staff has previously approved cask designs and issued technical specifications regarding a variety of items including, but not limited to, the following examples:

- general requirements and conditions regarding site-specific parameters, operating procedures, quality assurance, heavy loads, training, etc.

- a preoperational training exercise and demonstration of most cask operations, including loading, sealing, and drying (using mockups as appropriate); placement in storage; and return of fuel to the spent fuel pool

- specifications for the spent fuel to be stored in the cask, including, but not limited to, the type of spent fuel (i.e., BWR, PWR, or both), the maximum allowable enrichment of the fuel before irradiation, burnup (i.e., megawatt-days/MTU), the minimum acceptable cooling time of the spent fuel before storage in the cask, the maximum heat designed to be dissipated, the maximum spent fuel loading limit, the maximum neutron and gamma source terms, condition of the spent fuel (i.e., intact assembly or consolidated fuel rods, allowable cladding condition), and physical parameters (e.g., length, width, depth, weight, etc.)

- criticality controls,, such as cask water boron concentrations

- the inerting atmosphere requirements, such as vacuum drying and helium backfill parameters

- cask handling restrictions, such as lift height limits and ambient temperature (high/low) conditions

- confinement barrier requirements, such as leak rate limits

- thermal performance parameters, such as maximum temperatures or delta-temperatures

- radiological controls, such as radiation dose rates and contamination limits

- cask array and/or spacing limits for thermal performance and radiological considerations

Ensure that all necessary technical specifications are explicitly delineated in the SER and in the certificate of compliance or site-specific license, as applicable. These delineations typically restate the

Technical specifications defined in the SAR, but may be modified or supplemented as the staff deems appropriate.

NRC technical reviewers assigned to each SAR section have the responsibility to review the applicant's technical specifications and to identify any additional conditions that are necessary for cask use. The staff should ensure that the conditions for use, as evaluated and approved by the technical reviewers, complement one another and are not contradictory. The staff will coordinate the resolution of any disputed condition, limit, or specification. The staff is also responsible for identifying any unique specifications (e.g., administrative) that may not be covered in the technical sections, although input may be solicited from the technical reviewers regarding any topic.

Regulatory Guide 3.61[3] provides a recommended format for use by applicants in presenting Technical specifications. However, this format may not be applicable to all controls. Since the basis for the control may be extensively discussed in earlier sections of the SAR, the applicant may use an abbreviated format in SAR Section 12.

Coordinate the review of the proposed technical specifications with appropriate NRC staff familiar with ISFSI inspections and operations to assure the operational limitations are measurable and inspectable.

Upon completion of the review, the staff may prepare a separate table or appendix for SER Section 12 to explicitly designate the technical specifications that are applicable to the cask.

VI. Evaluation Findings

Review the applicable 10 CFR Part 72 acceptance criteria and provide a summary statement for each. These statements should be similar to the following model:

Table ___ of the safety evaluation report (SER) lists the technical specifications for the [cask designation]. These technical specifications are further discussed in Section 12 of the safety analysis report (SAR) [or other related document].

The staff concludes that the conditions for use of the [cask designation] identify necessary technical specifications to satisfy 10 CFR Part 72, and that the applicable acceptance criteria have been satisfied. The proposed technical specifications provide reasonable assurance that the cask will allow safe storage of spent fuel. This finding is reached on the basis of a review that considered the regulation itself, appropriate regulatory guides, applicable codes and standards, and accepted practices.

VII. References

1. *U.S. Code of Federal Regulations*, "Licensing Requirements for the Independent Storage of Spent Nuclear Fuel and High-level Radioactive Waste," Part 72, Title 10, "Energy."

2. *U.S. Code of Federal Regulations*, Part 20, "Standards for Protection Against Radiation," Title 10, "Energy."

3. U.S. Nuclear Regulatory Commission, "Standard Format and Content for a Topical Safety Analysis Report for a Spent Fuel Dry Storage Facility," Regulatory Guide 3.61, February 1989.

13.0 QUALITY ASSURANCE

I.   Review Objective

In this portion of the dry cask storage system (DCSS) review, the NRC evaluates the applicant's proposed quality assurance (QA) program, as described in the safety analysis report (SAR). In conducting this evaluation, the NRC staff seeks to ensure that the program provides adequate control over all activities related to the design, fabrication, assembly, testing, and use of DCSS structures, systems, and components (SSCs) that are important to safety.

To assess "adequate control," the staff determines whether the applicant's proposed QA program defines and assigns specific quality measures and controls to the various activities and SSCs. Moreover, the applicant should apply these quality measures and controls using a graded approach. The graded approach is described in NUREG/CR-6407[1]. That is, the effort expended on an activity or SSC should be consistent with its importance to safety. The QA program description provided in the SAR must identify both the procedures that the applicant will use to implement the QA program, as well as the activities and DCSS SSCs that are important to safety.

This evaluation should yield reasonable assurance that the applicant's proposed QA program will ensure that the DCSS will perform its intended functions in a satisfactory manner.

II. Areas of Review

This chapter of the DCSS Standard Review Plan (SRP) provides guidance for use in evaluating the applicant's proposed QA program. As described in Section V, "Review Procedures," a comprehensive evaluation involves examining the QA program in terms of the 18 criteria defined in 10 CFR Part 72[2], Subpart G, "Quality Assurance." Reviewers should obtain reasonable assurance that the applicant has implemented accepted QA principles in the design, fabrication, assembly, testing, and use of the DCSS SSCs. In addition, the SAR should address the assignment of specific QA levels to each activity and SSC important to safety.

It is essential that the SAR provide sufficient detail to enable the NRC staff to assess the adequacy of the proposed QA program. In addition, since many of the QA program controls may be detailed in other sections of the SAR, the description of the QA program in SAR Section 13 should reference other sections that contain relevant information. The QA program evaluation should therefore be coordinated with other aspects of the DCSS review. Such coordination will allow reviewers to derive a more accurate and complete assessment of the applicant's level of commitment to the overall QA program, the selection of quality criteria and quality levels, and the proposed implementation methods.

To control activities related to the design and development of the DCSS, the applicant must *first* establish and implement an effective design control program and associated QA program controls and implementing procedures. Consequently, in conducting the QA program evaluation, reviewers should emphasize the area of design control. An effective design control program will provide assurance that the proposed DCSS will be correctly designed and tested and will perform its intended function.

III. Regulatory Requirements

According to 10 CFR 72.24, "Contents of Application: Technical Information," the application must include, at a minimum, a description that satisfies the requirements of 10 CFR Part 72, Subpart G, "Quality Assurance," with regard to the QA program to be applied to the design, fabrication, construction, testing, and operation of the DCSS SSCs important to safety. Moreover, Subpart G states that the licensee shall establish the QA program at the earliest practicable time consistent with the schedule for accomplishing the activities.

IV. Acceptance Criteria

The SAR must provide a comprehensive description of the QA program that the applicant will establish, maintain, and execute to ensure control of all activities related to the design, fabrication, construction, testing, and use of DCSS SSCs that are important to safety. In addition, the QA program description in the SAR must identify the SSCs that are important to safety and must include information pertaining to managerial and administrative controls to ensure safe operation of the DCSS.

To be acceptable, the applicant should structure the QA program to apply QA measures and controls to all activities and SSCs using a graded approach in proportion to their importance to safety. Consequently, in identifying the activities and SSCs that are important to safety, the SAR should also identify the associated degree of importance for each. Those activities and SSCs that are highly important to safety should be covered by a high level of control, while those less important to safety may have a lower level of control.

An applicant may choose to apply the highest level of QA and control to all activities and SSCs without distinction. By contrast, a graded approach to QA requires applicant justification and reviewer acceptance. Consequently, the SAR should adequately describe the proposed graded approach.

Section V summarizes the acceptance positions that the NRC staff uses to evaluate an applicant's QA program as described in the SAR. These positions represent solutions and approaches that the staff finds acceptable, but they may not be the only possible solutions and approaches. Various alternatives to the detailed guidance in this SRP may be deemed acceptable, provided that the applicant adequately documents and justifies the deviations.

For each of the activities and SSCs identified as important to safety, the applicant should identify and define the level of control to be applied to each of the following 18 elements of the QA program. Appendix A presents a sample checklist for use in evaluating each of these QA elements. The NRC intends that the attributes listed in the Appendix for each element are to be applied collectively only in the most stringent application of the QA program. Lesser quality requirements may be effected by modifying or eliminating some attributes from selected elements. The applicant's QA program, and associated QA program controls and implementing procedures regarding activities performed, must be in place before activities begin.

1. Quality Assurance Organization

The SAR should describe (and illustrate in an appropriate chart) the organizational structure, interrelationships, and areas of functional responsibility and authority for all organizations performing quality- and safety-related activities, including both the applicant's organization and principal contractors, if applicable. Persons or organizations responsible for ensuring that an appropriate QA program has been established and verifying that activities affecting quality have been correctly performed should have sufficient authority, access to work areas, and organizational freedom to carry out that responsibility.

2. Quality Assurance Program

The SAR should provide acceptable evidence that the applicant's proposed QA program will be well-documented, planned, implemented, and maintained to provide the appropriate level of control over activities and SSCs, consistent with their relative importance to safety.

3. Design Control

The SAR should describe the approach that the applicant will use to define, control, and verify the design and development of the DCSS. An effective design control program will provide assurance that the proposed DCSS will be appropriately designed and tested and will perform its intended function.

4. Procurement Document Control

Documents used to procure SSCs or services should include or reference applicable design bases and other requirements necessary to ensure adequate quality. To the extent necessary, these procurement documents should require that suppliers have a QA program consistent with the quality level of the SSCs or services to be procured.

5. Instructions, Procedures, and Drawings

The SAR should define the applicant's proposed procedures for ensuring that activities affecting quality will be prescribed by, and performed in accordance with, documented instructions, procedures, or drawings of a type appropriate for the circumstances.

6. Document Control

The SAR should define the applicant's proposed procedures for preparing, issuing, and revising documents that specify quality requirements or prescribe activities affecting quality. These procedures should provide adequate control to ensure that only the latest documents are used. In addition, the applicant's authorized personnel should carefully review and approve the accuracy of all documents and associated revisions before they are released for use.

7. Control of Purchased Material, Equipment, and Services

The SAR should define the applicant's proposed procedures for controlling purchased material, equipment, and services to ensure conformance with specified requirements.

8. Identification and Control of Materials, Parts, and Components

The SAR should define the applicant's proposed provisions for identifying and controlling materials, parts, and components to ensure that incorrect or defective SSCs are not used.

9. Control of Special Processes

The SAR should describe the controls that the applicant will establish to ensure the acceptability of special processes (such as welding, heat treatment, nondestructive testing, and chemical cleaning) and that they are performed by qualified personnel using qualified procedures and equipment.

10. Licensee Inspection

The SAR should define the applicant's proposed provisions for inspection of activities affecting quality to verify conformance with instructions, procedures, and drawings.

11. Test Control

The SAR should define the applicant's proposed provisions for tests to verify that SSCs conform to specified requirements and will perform satisfactorily in service. The applicant should specify test requirements in written procedures, including provisions for documenting and evaluating test results. In addition, the applicant should establish qualification programs for test personnel.

12. Control of Measuring and Test Equipment

The SAR should define the applicant's proposed provisions to ensure that tools, gauges, instruments, and other measuring and testing devices are properly identified, controlled, calibrated, and adjusted at specified intervals.

13. Handling, Storage, and Shipping Control

The SAR should define the applicant's proposed provisions to control the handling, storage, shipping, cleaning, and preservation of SSCs in accordance with work and inspection instructions to prevent damage, loss, and deterioration.

14. Inspection, Test, and Operating Status

The SAR should define the applicant's proposed provisions to control the inspection, test, and operating status of SSCs to prevent inadvertent use or bypassing of inspections and tests.

15. Nonconforming Materials, Parts, or Components

The SAR should define the applicant's proposed provisions to control the use or disposition of nonconforming materials, parts, or components.

16. Corrective Action

The SAR should define the applicant's proposed provisions to ensure that conditions adverse to quality are promptly identified and corrected and that measures are taken to preclude recurrence.

17. Quality Assurance Records

The SAR should define the applicant's proposed provisions for identifying, retaining, retrieving, and maintaining records that document evidence of the control of quality for activities and SSCs important to safety.

18. Audits

The SAR should define the applicant's proposed provisions for planning, scheduling, and conducting audits to verify compliance with all aspects of the QA program, and to determine the effectiveness of the overall program. The SAR should clearly identify responsibilities and procedures for conducting audits, documenting and reviewing audit results, and designating management levels to review and assess audit results. In addition, the SAR should describe the applicant's provisions for incorporating the status of audit recommendations in management reports.

V.   Review Procedures

Except in cases in which the applicant proposes an acceptable alternative method for complying with specified portions of the Commission's regulations, NRC staff reviewers should use the following methods and procedures to evaluate conformance of the applicant's proposed QA program with the Commission's regulations.

Begin the QA program review by determining whether the application includes the information required by 10 CFR Part 72 with regard to the QA program, as well as the topics discussed in this chapter of the DCSS SRP. If deficiencies are identified in the application, request that the applicant submit additional information before proceeding further with the QA program review.

Next, review each element of the QA program description with respect to the acceptance criteria in Section IV, above. The primary objective of this review is to ensure that the applicant has provided sufficient information to support a reviewer's conclusion that the proposed QA program meets the stated acceptance criteria. Consequently, the review should rely exclusively on an assessment of the information presented.

Determine whether the applicant has adequately planned the work to be accomplished, and whether the necessary policies, procedures, and instructions either are in place or will be in place before work begins. This review should provide reasonable assurance that adequate coordination exists among the applicant's QA, configuration management, and maintenance programs, and that the QA program is an integral part of everyday work activities. In addition, this review should provide reasonable assurance that the applicant will be able to monitor the effectiveness of the implementation of the QA program and will make needed adjustments on a timely basis.

Look beyond the existence of appropriate elements, and assess the effectiveness of the applicant's QA program design. Determine whether the applicant's QA program addresses the full scope of the application. The QA program description should specify the QA criteria, the basis on which the criteria were selected, how the criteria are apportioned within the sections of the application, and the proposed implementation method for each.

Write the related sections of the safety evaluation report (SER) summarizing the conduct of the review. Identify the material in the application that forms the basis for the finding of reasonable assurance with respect to the acceptance criteria, and present any recommendations for modification of the application that might be necessary to allow a finding of reasonable assurance.

VI. Evaluation Findings

The staff's review should verify that the application provided sufficient information to facilitate a thorough and comprehensive evaluation of the applicant's QA program. This information should demonstrate that the applicant has conceived and implemented adequate program provisions to ensure the quality of management control over procedures, processes, and SSCs important to the health and safety of workers and the public and the protection of the environment. In addition, the information provided in the application should be consistent with the guidance in this SRP and the related regulatory requirements.

The review record should demonstrate that the staff has reviewed the applicant's QA program for the design, fabrication, construction, testing, and operation of the DCSS, according to the guidance in this chapter of the SRP. In addition, the record should state the staff's finding that the application provided sufficient information, and that the review is sufficiently complete to support the following conclusions (in either the SER or a letter to the applicant):

On the basis of the staff's detailed review and evaluation of the QA program described in the [topical report or safety analysis report (SAR)] for the [DCSS name], the NRC staff has reached the following conclusions:

1. The structure of the organization and the assignment of responsibility for each activity ensure that designated responsible parties will perform the necessary work to achieve and maintain the specified quality requirements. Conformance to established requirements will be verified by individuals and groups not directly responsible for performing the work. The organizations responsible for verifying quality report through a management hierarchy that allows the required authority and organizational freedom, including sufficient independence from influences of cost and schedule.

2. The QA program is well-documented and provides adequate control over activities affecting quality, as well as structures, systems, and components (SSCs) that are important to safety, to the extent consistent with their relative importance to safety. The QA program describes a management system and controls that, when properly implemented, will comply with the requirements of Subpart G to 10 CFR Part 72 and 10 CFR Part 21[3].

3. Accordingly, the staff concludes that the applicant's QA program complies with the applicable NRC regulations and industry standards and can be implemented for the [specify the application].

In addition, the SER should provide a brief description of the applicant's QA program, with highlights of the more important aspects of the program.

VII. References

1. NUREG/CR-6407, "Classification of Transportation Packaging and Dry Spent Fuel Storage System Components According to Importance to Safety," February 1996.

2. *U.S. Code of Federal Regulations*, Part 72, "Licensing Requirements for the Independent Storage of Spent Nuclear Fuel and High-Level Radioactive Waste," Title 10, "Energy,"

3. *U.S. Code of Federal Regulations*, Part 21, "Reporting of Defects and Noncompliance" Title 10, "Energy,".

14.0 DECOMMISSIONING

I.   Review Objective

The decommissioning review ensures that the safety analysis report (SAR) demonstrates that the applicant has conceived adequate provisions to facilitate transfer of the spent fuel stored in the ISFSI to the DOE and provide for the future decontamination and disposal of the components that make up the dry cask storage system (DCSS).

The NRC recognizes that decommissioning will occur in the distant future (perhaps more than 20 years after the cask is first used) and will employ site-specific procedures available at that time. Consequently, 10 CFR Part 72[1] does not require licensees to develop detailed decommissioning plans until near the time of license termination.

By contrast, during the licensing of a proposed Dry Cask Storage System (DCSS), the applicant need only submit a *conceptual* decommissioning plan for NRC evaluation. Nonetheless, the applicant's conceptual plan must provide reasonable assurance that the owner of the DCSS can conduct decontamination and decommissioning in a manner that adequately protects the health and safety of the public.

Specifically, the conceptual decommissioning plan must identify the types of waste to be generated, the anticipated types of contamination, the proposed practices and procedures for decontaminating the cask and disposing of residual radioactive materials.

To augment the conceptual plan, the NRC requires a commitment that *general* licensees submit a detailed plan for ISFSI decommissioning along with their reactor decommissioning plan. Similarly, *site-specific* licensees will submit a detailed decommissioning plan for review and approval before initiating decommissioning activities at the facility.

II.   Areas of Review

This portion of the DCSS review evaluates the applicant's conceptual decommissioning plan to ensure that it provides reasonable assurance that the licensee can conduct decontamination and decommissioning in a manner that adequately protects the health and safety of the public. Consequently, this chapter of the DCSS Standard Review Plan (SRP) provides guidance for use in conducting a comprehensive evaluation of the conceptual plan, which *may* encompass the following areas of review, as described in Section V, "Review Procedures":

1.   identification and discussion of the anticipated decommissioning activities, types of waste to be generated, possible types of contamination, and planned waste disposal method(s)

2.   commitment to decontaminate the facility to applicable NRC criteria

3.   a financial plan, providing adequate financial assurance for the cost of decommissioning, submitted as a separate document, as required by Regulatory Guide (RG) 3.50[2]

4.   commitment to submit a timely, detailed decommissioning plan for NRC review and approval before initiating decommissioning activities

III.   Regulatory Requirements

The requirements applicable to this portion of the DCSS review represent the following four distinct areas:

1.   The ISFSI or MRS must be designed for decommissioning. Provisions must be made to facilitate decontamination of structures and equipment, minimize the quantity of radioactive wastes and contaminated equipment, and facilitate the removal of radioactive wastes and contaminated materials at the time the ISFSI or MRS is permanently decommissioned. [10 CFR 72.130.]

2. The cask must be designed to facilitate decontamination to the extent practicable. [10 CFR 72.236(i).]

3. The requirements for financial assurance and record keeping associated with decommissioning are found in 10 CFR 72.30.

4. The requirements for terminating an ISFSI license and decommissioning ISFSI sites and buildings are found in including the requirements for submitting the final decommissioning plan are found in [10 CFR 72.54.]

IV. Acceptance Criteria

In general, the DCSS decommissioning evaluation seeks to ensure that the given design and conceptual decommissioning plan fulfill the following acceptance criteria:

1. decontamination of buildings and equipment, as specified in RG 1.86[3].

2. classification and disposal of wastes, as contained in 10 CFR 61.55[4].

V. Review Procedures

Review the general description and operating features of the cask system and its application to an ISFSI facility, as presented in SAR Section 1. Then, review the conceptual decommissioning plan in SAR Section 2. In particular, verify that the applicant has accurately and acceptably identified (1) the types of waste to be generated, (2) the anticipated types of contamination, (3) the proposed practices and procedures for decontaminating the cask and disposing of residual radioactive materials after all spent fuel and spent fuel casks have been removed from the site, and (4) the projected decommissioning activities. Note that the final acceptance criteria will be measured and evaluated at the time of decommissioning, and further guidance will be added to this SRP as it becomes available.

The NRC accepts that cask system features required to provide other capabilities may counter or interfere with features intended to facilitate decommissioning. For example, use of steel to provide strength may result in greater activation of materials; structural integrity designed to provide required safety in design-basis events increases the difficulty associated with demolition and size reduction, as does the increased material mass used for radiation and physical shielding. The NRC has accepted the priority of safety-related features and capabilities over decommissioning considerations when such trade-offs arise.

After the casks have been decontaminated, the major radiation sources may be those resulting from activation of the system components, such as the concrete shielding and reinforcing steel. Verify that the applicant has properly estimated the activities of these nuclides. Several activation products are short-lived, and the SAR may discuss their activity as a function of time after unloading. Although the specific activation products depend on the materials initially present in the cask components, the nuclides of interest are generally Cr-51, Mn-55, Fe-58, Co-58, Co-60, and Ni-63. A significant reduction in the total activation occurs in only 1 year after unloading.

Because of the low levels of spontaneous fission and subcritical multiplication in the spent fuel during the storage period, the activation of the cask components is generally very minor and can be approximated by simple, conservative methods. A typical approach is to use the same flux calculated from the deterministic shielding analysis (documented in SAR Section 5), along with appropriate cross-sections from the same calculation. For conservatism, activation of the cask body is determined from the flux at the inner surface. Equilibrium activities of the irradiated structures are generally calculated without considering the time dependence of the flux during the storage period.

Another common approach is to determine a conservative one-group flux, select a conservative cross-section library, and calculate the activities of the resulting radioactive nuclides using ORIGEN2[5]. This evaluation is intended to ensure that the activated cask components can be disposed of in a low-level waste disposal site. The acceptable criteria are specified in the tables contained in 10 CFR 61.55[2]. Radiological survey information in the applicant's final decommissioning plan should verify the preliminary waste classification and the acceptability of the proposed disposal methods. Other wastes generated from the applicant's decontamination activities should also be reviewed and evaluated to determine their quantity and acceptability for disposal under 10 CFR Part 61.

VI. Evaluation Findings

Review the 10 CFR Part 72 acceptance criteria and provide a summary statement for each. These statements should be similar to the following model:

The applicant's proposed cask design includes adequate provisions for decontamination and decommissioning. As discussed in Section ___ of the SAR, these provisions include facilitating decontamination of the DCSS, if needed; storing the remaining components, if no waste facility is expected to be available; and disposing of any remaining low-level radioactive waste.

Section ___ of the SAR presents information concerning the proposed practices and procedures for decontaminating the cask and disposing of residual radioactive materials after all spent fuel has been removed. This information provides reasonable assurance that the applicant will conduct decontamination and decommissioning in a manner that adequately protects the health and safety of the public.

The staff concludes that the decommissioning considerations for the [cask designation] are in compliance with 10 CFR Part 72. This evaluation provides reasonable assurance that the [cask designation] will allow safe storage of spent fuel.This finding is reached on the basis of a review that considered the regulation itself, appropriate regulatory guides, applicable codes and standards, and accepted engineering practices.

VII. References

1. *U.S. Code of Federal Regulations*, "Licensing Requirements for the Independent Storage of Spent Nuclear Fuel and High-level Radioactive Waste," Part 72, Title 10, "Energy."

2. U.S. Nuclear Regulatory Commission, Standard Format and Content for a License Application to Store Spent Fuel and High-Level Radioactive Waste," Regulatory Guide 3.50, September 1989.

3. U.S. Nuclear Regulatory Commission,"Termination of Operating Licenses for Nuclear Reactors," Regulatory Guide 1.86, June 1974.

4. *U.S. Code of Federal Regulations*, Part 61, "Licensing Requirements for Land Disposal of Radioactive Wastes," Title 10, "Energy."

5. Oak Ridge National Laboratory, "ORIGEN2: Isotope Generation and Depletion Code-Matrix Exponential Method,", 1991.

APPENDIX A
SAMPLE CHECKLIST FOR EVALUATING QUALITY ASSURANCE
PROGRAM ELEMENTS FOR DRY CASK STORAGE SYSTEMS

Within the context of "Licensing Requirements for the Independent Storage of Spent Nuclear Fuel and High-Level Radioactive Waste," Subpart G of 10 CFR Part 72 specifies 18 criteria that must be addressed in a quality assurance (QA) program. To be found acceptable, the applicant's proposed QA program should include provisions for meeting each of the following acceptance criteria:

1. Quality Assurance Organization

The SAR should describe the structure, interrelationships, and areas of functional responsibility and authority for all organizational elements that will perform activities related to quality and safety. Acceptability of these organizational elements is contingent upon the following criteria:

a. The applicant should retain and exercise responsibility for the QA program. The assignment of responsibility for the overall QA program in no degree relieves line management of their responsibility for the achievement of quality.

b. The application should identify and describe the QA functions, performed by the applicant's QA organization or delegated to other organizations, that will provide controls to ensure implementation of the applicable elements of the QA criteria.

c. Clear management controls and effective lines of communication should exist between the applicant's QA organizations and suppliers to ensure proper direction of the QA program and resolution of QA problems.

d. Organization charts should identify onsite and offsite organizational elements that will function under the purview of the QA program and the lines of responsibility.

e. High-level management should be responsible for documenting and promulgating the applicant's QA policies, goals, and objectives, and this management level should maintain a continuing involvement in QA matters. The application should also describe the lines of communication between intermediate levels of management and between this position and the Manager (or Director) of QA.

f. The applicant should designate a position that retains overall authority and responsibility for the QA program.

g. The authority and independence of the individual responsible for managing the QA program should be such that he or she can direct and control the organization's QA program, can effectively ensure conformance to quality requirements, and can remain sufficiently independent of undue influences and responsibilities of schedules and costs. An acceptable organizational structure would have this individual report to at least the same organizational level as the highest line manager directly responsible for performing activities affecting quality.

h. Individuals or groups responsible for defining and controlling the content of the QA program and related manuals should have appropriate organizational position and authority, as should the management level responsible for final review and approval.

I. The qualification requirements for the principal QA management positions should demonstrate management and technical competence commensurate with the responsibilities of these positions.

j. Conformance to established requirements should be verified by individuals or groups who do not have direct responsibility for performing the work being verified. The quality control function may be part of the line organization, provided that the QA organization performs periodic surveillance to confirm sufficient independence from the individuals who performed the activities.

k. Persons and organizations performing QA functions should have direct access to management levels that will ensure accomplishment of quality-affecting activities. These individuals should have sufficient authority and organizational freedom to perform their QA functions effectively and without reservation. In addition, they should be able to identify quality problems; initiate, recommend, or provide solutions through designated channels; and verify implementation of solutions.

l. Designated QA individuals or organizations should have the responsibility and authority, delineated in writing, to stop unsatisfactory work and control further processing, delivery, or installation of nonconforming material. In addition, the application should describe how stop-work requests will be initiated and completed.

m. The extent of QA controls should be determined by the QA staff in combination with the line staff and should depend upon the specific activity or item complexity and level of importance to safety.

2. Quality Assurance Program

The SAR should provide acceptable evidence that the applicant's proposed QA program will be well-documented, planned, implemented, and maintained to provide the appropriate level of control over activities and SSCs, consistent with their relative importance to safety. Acceptability of the QA program description is contingent upon the following criteria:

a. The applicant should specify measures used to ensure that the QA program meets applicable acceptance criteria.

b. Management should commit to regularly assess the effectiveness of the QA program. In addition, the applicant should describe how management (above and beyond the QA organization) will regularly assess the scope, status, adequacy, and compliance of the QA program to the requirements of 10 CFR Part 72. These measures should include frequent contact with program status through reports, meetings, and audits, as well as performance of a periodic assessment that is planned and documented with corrective action identified and tracked.

c. The applicant should specify measures used to ensure that trained, qualified personnel within the organization will be assigned to determine that functions delegated to contractors are properly accomplished.

d. The applicant should briefly summarize the corporate QA policies, goals, and objectives and should establish a meaningful channel for transmittal of these policies, goals, and objectives down through the levels of management.

e. The applicant should designate responsibilities for implementing the major activities addressed in the QA manuals.

f. The applicant should establish provisions to control the distribution of the QA manuals and revisions.

g. The applicant should establish provisions for communicating to all responsible organizations and individuals that policies, QA manuals, and procedures are mandatory requirements.

h. The applicant should provide a comprehensive listing of QA procedures, plus a matrix of these procedures cross-referenced to each of the QA criteria, to demonstrate that the QA program will be fully implemented by documented procedures.

I. The applicant should identify the structures, systems, and components (SSCs) that are important to safety and therefore will be controlled by the QA program.

j. The applicant should review and document agreement with the QA program provisions of its suppliers to ensure implementation of a program meeting the QA criteria.

k.   The applicant should establish provisions for the resolution of disputes involving quality, arising from a difference of opinion between QA/QC personnel and personnel from other departments (engineering, procurement, manufacturing, etc.).

l.   The applicant should establish indoctrination, training, and qualification programs that will fulfill the following criteria:

   ● Personnel responsible for performing activities affecting quality should be instructed as to the purpose, scope, and implementation of the quality-related manuals, instructions, and procedures.

   ● Personnel performing activities affecting quality should be trained and qualified in the principles and techniques of the activities being performed.

   ● The applicant should maintain the proficiency of personnel performing quality-affecting activities by retraining, reexamining, and recertifying.

   ● The applicant should describe specific documentation of completed training and qualification.

   ● Qualified personnel should be certified in accordance with accepted codes and standards.

3. Design Control

The SAR should describe the approach that the applicant will use to define, control, and verify the design and development of the dry cask storage system (DCSS). Acceptability of activities related to design control is contingent upon the following criteria:

a.   The applicant should establish measures to carry out design activities in a planned, controlled, and orderly manner.

b.   The applicant should establish measures to correctly translate the applicable regulatory requirements and design bases into specifications, drawings, written procedures, and instructions.

c.   The applicant should specify quality standards in the design documents, and should control deviations and changes from these quality standards.

d.   The applicant should review designs to ensure that design characteristics can be controlled, inspected, and tested and that inspection and test criteria are identified.

e.   The applicant should establish both internal and external design interface controls. These controls should include review, approval, release, distribution, and revision of documents involving design interfaces with participating design organizations.

f.   The applicant should properly select and perform design verification processes, such as design reviews, alternative calculations, or qualification testing. When a test program is to be used to verify the adequacy of a design, the applicant should use a qualification test of a prototype unit under adverse design conditions.

g.   Design verification constitutes confirmation that the design of the SSC is suitable for its intended purpose. Consequently, design verification requires a level of skill at least equal to that of the original designer, while design checking can be performed by a less experienced person. (As an example, design checking, which must also be performed, includes confirmation of the numerical accuracy of computations and the accuracy of data input to computer codes. Confirmation that the correct computer code has been used is part of design verification.) · Design verification should be performed by persons other than those performing design checking. In addition, individuals or groups responsible for design verification should not include the original designer, and normally should not include the designer's immediate supervisor.

h. Design and specification changes are subject to the same design controls and the same or equivalent approvals that were applicable to the original design.

I. The applicant should document all errors and deficiencies in the design, or the design process, that could adversely affect SSCs important to safety. In addition, the applicant should take adequate corrective action, including root cause evaluation of significant errors and deficiencies, to preclude repetition.

j. Before selecting materials, parts, and equipment that are standard, commercial (off-the-shelf), or have been previously approved for a different application, the applicant should review their suitability for the intended application.

k. The applicant should implement written procedures to identify and control the authority and responsibilities of all individuals or groups responsible for design reviews and other design verification activities.

l. The applicant should establish measures that include the use of valid industry standards and specifications for the selection of suitable materials, parts, equipment, and processes for SSCs that are important to safety.

4. Procurement Document Control

Documents used to procure SSCs or services should include or reference applicable design bases and other requirements necessary to ensure adequate quality. Acceptability of the proposed procurement document controls is contingent upon the following criteria:

a. The applicant should establish procedures that clearly delineate the sequence of actions to be accomplished in the preparation, review, approval, and control of procurement documents.

b. Qualified personnel should review and concur with the adequacy of quality requirements stated in procurement documents. This review should ensure that the quality requirements are correctly stated, inspectable, and controllable; there are adequate acceptance and rejection criteria; and the procurement document has been prepared, reviewed, and approved in accordance with QA program requirements.

c. The applicant should document the review and approval of procurement documents before they are released, and the documentation should be available for verification.

d. Procurement documents should identify the applicable QA requirements that must be compiled with and described in the supplier's QA program. In addition, the applicant should review and concur with the supplier's QA program.

e. Procurement documents should contain or reference the regulatory requirements, design bases, and other technical requirements.

f. Procurement documents should identify the documentation (e.g., drawings, specifications, procedures, inspection and fabrication plans, inspection and test records, personnel and procedure qualifications, and chemical and physical test results of material) to be prepared, maintained, and submitted to the purchaser for review and approval.

g. Procurement documents should identify records to be retained,. controlled, and maintained by the supplier, and those to be delivered to the purchaser before use or installation of the hardware.

h. Procurement documents should specify the procuring agency's right of access to the supplier's facilities and records for source inspection and audit.

i. Changes and revisions to procurement documents should be subject to the same or equivalent review and approval as the original documents.

5. Instructions, Procedures, and Drawings

The SAR should define the applicant's proposed procedures for ensuring that activities affecting quality will be prescribed by, and performed in accordance with, documented instructions, procedures, or drawings of a type appropriate for the circumstances. Acceptability of the proposed instructions, procedures, or drawings is contingent upon the following criteria:

a. Activities affecting quality should be prescribed and accomplished in accordance with documented instructions, procedures, or drawings.

b. The applicant should establish provisions that clearly delineate the sequence of actions to be accomplished in the preparation, review, approval, and control of instructions, procedures, and drawings.

c. The instructions, procedures, and drawings should specify the methods for complying with each of the applicable QA criteria.

d. Instructions, procedures, and drawings should include quantitative acceptance criteria (such as dimensions, tolerances, and operating limits), as well as qualitative acceptance criteria (such as workmanship samples), as verification that activities important to safety have been satisfactorily accomplished.

e. The QA organization should review and concur with the procedures, drawings and specifications related to inspection plans, tests, calibrations, and special processes, as well as any subsequent changes to these documents.

6. Document Control

The SAR should define the applicant's proposed procedures for preparing, issuing, and revising documents that specify quality requirements or prescribe activities affecting quality. Acceptability of the proposed document control procedures is contingent upon the following criteria:

a. The application should identify all documents to be controlled under this subsection. As a minimum this should include design specifications; design and fabrication drawings; procurement documents; QA manuals; design criteria documents; fabrication, inspection, and testing instructions; and test procedures.

b. The applicant should establish procedures to control the review, approval, and issuance of documents and changes thereto, before release, to ensure that the documents are adequate and applicable quality requirements are stated.

c. The applicant should establish provisions to identify individuals or groups responsible for reviewing, approving, and issuing documents and revisions thereto.

d. Document revisions should be reviewed and approved by the same organizations that performed the original review and approval or by other qualified responsible organizations designated by the applicant.

e. Approved changes should be included in instructions, procedures, drawings, and other documents before the change is implemented.

f. The applicant should control obsolete or superseded documents to prevent inadvertent use.

g. Documents should be available at the location where the activity is performed.

h. The applicant should establish a master list (or equivalent) to identify the current revision number of instructions, procedures, specifications, drawings, and procurement documents. In addition, the applicant should update this list and distribute it to predetermined, responsible personnel to preclude use of superseded documents.

7. Control of Purchased Material, Equipment, and Services

The SAR should define the applicant's proposed procedures for controlling purchased material, equipment, and services to ensure conformance with specified requirements. Acceptability of the proposed controls is contingent upon the following criteria:

a. Qualified personnel should evaluate the supplier's capability to provide services and products of acceptable quality before the award of the procurement order or contract. The applicant's QA and engineering groups should participate in the evaluation of those suppliers providing critical items and services important to safety, and the applicant should define the responsibilities for each group's participation.

b. The applicant should evaluate suppliers on the basis of one or more of the following criteria:

   - the supplier's capability to comply with the elements of the QA criteria that are applicable to the type of material, equipment, or service being procured

   - review of previous records and performance of suppliers who have provided similar articles or services of the type being procured

   - a survey of the supplier's facilities and QA program to assess the capability to supply a product that meets applicable design, manufacturing, and quality requirements

c. The applicant should document and file the results of supplier evaluations.

d. The applicant should plan and perform adequate surveillance of suppliers during fabrication, inspection, testing, and shipment of materials, equipment, and components in accordance with written procedures to ensure conformance to the purchase order requirements. These procedures provide the following information:

   - instructions that specify the characteristics or processes to be witnessed, inspected or verified, and accepted; the method of surveillance and the extent of documentation required; and those responsible for implementing these instructions

   - procedures for audits and surveillance to ensure that the supplier complies with the quality requirements (surveillance should be performed for SSCs for which verification of procurement requirements cannot be determined upon receipt)

e. As a minimum, the supplier should furnish the following records to the purchaser:

   - documentation that identifies the purchased material or equipment and the specific procurement requirements (e.g., codes, standards, and specifications) met by the items

   - documentation that identifies any procurement requirements that have not been met, together with a description of any nonconformances designated "accept as is" or "repair"

The applicant should describe the proposed procedures for reviewing and accepting these documents and, as a minimum, should ensure that this review and acceptance will be undertaken by a responsible QA individual.

f. The applicant should conduct periodic audits, independent inspections, or tests to ensure the validity of the suppliers' certificates of conformance.

g. The applicant should perform a receiving inspection of the supplier-furnished material, equipment, and services to ensure fulfillment of the following criteria:

   - The material, component, or equipment should be properly identified in a manner that corresponds with the identification on the purchasing and receiving documentation.

- Material, components, equipment, and acceptance records should be inspected and judged acceptable in accordance with predetermined inspection instructions before installation or use.

- Inspection records or certificates of conformance attesting to the acceptance of material, components, and equipment should be available before installation or use.

- Items accepted and released should be identified as to their inspection status before they are forwarded to a controlled storage area or released for installation or further work.

  h.   The applicant should assess the effectiveness of suppliers' quality controls at intervals consistent with the importance to safety, complexity, and quantity of the SSCs procured.

8. **Identification and Control of Materials, Parts, and Components**

    The SAR should define the applicant's proposed provisions for identifying and controlling materials, parts, and components to ensure that incorrect or defective SSCs are not used. Acceptability of the proposed controls is contingent upon the following criteria:

    a.   The applicant should establish procedures to identify and control materials, parts, and components (including partially fabricated subassemblies).

    b.   The applicant should determine identification requirements during generation of specifications and design drawings.

    c.   The identification and control procedures should ensure that identification will be maintained either on the item or on records traceable to the item to preclude use of incorrect or defective items.

    d.   Identification of materials and parts of important-to-safety items should be traceable to the appropriate documentation (such as drawings, specifications, purchase orders, manufacturing and inspection documents, deviation reports, and physical and chemical mill test reports).

    e.   The location and method of identification should not affect the fit, function, or quality of the item being identified.

    f.   The applicant should verify and document the correct identification of all materials, parts, and components before releasing them for fabrication, assembly, shipping, and installation.

9. Control of Special Processes

    The SAR should describe the controls that the applicant will establish to ensure the acceptability of special processes (such as welding, heat treatment, nondestructive testing, and chemical cleaning) and that they are performed by qualified personnel using qualified procedures and equipment. Acceptability of the proposed controls is contingent upon the following criteria:

    a.   The applicant should establish procedures to control special processes (such as welding, heat treating, nondestructive testing, and cleaning), for which direct inspection is generally impossible or disadvantageous. In addition, the applicant should provide a listing of these special processes.

    b.   The applicant should qualify procedures, equipment, and personnel connected with special processes, in accordance with applicable codes, standards, and specifications.

    c.   Qualified personnel should perform special processes in accordance with written process sheets (or the equivalent) with recorded evidence of verification.

    d.   The applicant should establish, file, and keep current qualification records of procedures, equipment, and personnel associated with special processes.

10. Licensee Inspection

The SAR should define the applicant's proposed provisions for inspection of activities affecting quality to verify conformance with instructions, procedures, and drawings. Acceptability of the proposed provisions is contingent upon the following criteria:

a. The applicant should establish, document, and conduct an inspection program that effectively verifies conformance of quality-affecting activities with requirements in accordance with written, controlled procedures.

b. Inspection personnel should be sufficiently independent from the individuals performing the activities being inspected.

c. Inspection procedures, instructions, and check lists should provide the following details:

- identification of characteristics and activities to be inspected

- identification of the individuals or groups responsible for performing the inspection operation

- acceptance and rejection criteria

- a description of the method of inspection

- procedures for recording evidence of completing and verifying a manufacturing, inspection, or test operation

- identification of the recording inspector or data recorder and the results of the inspection operation

d. The applicant should use inspection procedures or instructions with the necessary drawings and specifications when performing inspection operations.

e. The applicant should qualify inspectors in accordance with applicable codes, standards, and company training programs. Their qualifications and certifications should be kept current.

f. The applicant should inspect modifications, repairs, and replacements in accordance with the original design and inspection requirements or acceptable alternatives.

g. The applicant should establish provisions that identify mandatory inspection hold points for witnessing by a designated inspector.

h. The applicant should identify the individuals or groups who will perform receiving and process verification inspections, and should demonstrate that they have sufficient independence and qualifications.

i. The applicant should establish provisions for indirect control by monitoring processing methods, equipment, and personnel if direct inspection is not possible.

11. Test Control

The SAR should define the applicant's proposed provisions for tests to verify that SSCs conform to specified requirements and will perform satisfactorily in service. Acceptability of the proposed provisions is contingent upon the following criteria:

a. The applicant should establish, document, and conduct a test program to demonstrate that the item will perform satisfactorily in service in accordance with written, controlled procedures.

b. Written test procedures should incorporate or reference the following information:

- requirements and acceptance limits contained in applicable design and procurement documents

- instructions for performing the test

- test prerequisites

- mandatory inspection hold points

- acceptance and rejection criteria

- methods of documenting or recording test data results

c. A qualified, responsible individual or group should document test results and evaluate their acceptability. When practicable, the applicant should test the SSC under conditions that will be present during normal and anticipated off-normal operations.

12. Control of Measuring and Test Equipment

The SAR should define the applicant's proposed provisions to ensure that tools, gauges, instruments, and other measuring and testing devices are properly identified, controlled, calibrated, and adjusted at specified intervals. Acceptability of the proposed provisions is contingent upon the following criteria:

a. Documented procedures should describe the calibration technique and frequency, maintenance, and control of all measuring and test equipment (instruments, tools, gauges, fixtures, reference and transfer standards, and nondestructive test equipment) that will be used in the measurement, inspection, and monitoring of SSCs that are important to safety.

b. Measuring and test equipment should be identified and traceable to the calibration test data.

c. The applicant should label, tag, or otherwise document measuring and test equipment to indicate the date of the next scheduled calibration and to provide traceability to calibration test data.

d. The applicant should calibrate measuring and test instruments at specified intervals on the basis of the required accuracy, precision, purpose, degree of usage, stability characteristics, and other conditions that could affect the accuracy of the measurements.

e. When measuring and test equipment is found to be out of calibration, the applicant should take and document measures to assess the validity of previous inspections.

f. The applicant should document and maintain the complete status of all items under the calibration system.

g. Reference and transfer standards should be traceable to nationally recognized standards; where national standards do not exist, the applicant should establish provisions to document the basis for calibration.

13. **Handling, Storage, and Shipping Control**

The SAR should define the applicant's proposed provisions to control the handling, storage, shipping, cleaning, and preservation of SSCs in accordance with work and inspection instructions to prevent damage, loss, and deterioration. Acceptability of the proposed provisions is contingent upon the following criteria:

a. Qualified individuals should establish and accomplish special handling, preservation, storage, cleaning, packaging, and shipping requirements in accordance with predetermined work and inspection instructions.

b. The applicant should prepare procedures to control the cleaning, handling, storage, packaging, shipping, and preservation of materials, components, and systems in accordance with design and specification requirements to preclude damage, loss, or deterioration by environmental conditions (such as temperature or humidity).

14. Inspection, Test, and Operating Status

The SAR should define the applicant's proposed provisions to control the inspection, test, and operating status of SSCs to prevent inadvertent use or bypassing of inspections and tests. Acceptability of the proposed provisions is contingent upon the following criteria:

a. The applicant should know the inspection and test status of items throughout fabrication.

b. Established procedures should control the application and removal of inspection and welding stamps and operating status indicators (such as tags, markings, labels, and stamps).

c. Procedures under the cognizance of the QA organization should control bypasses of required inspections, tests, and other critical operations.

d. The applicant should specify the organization responsible for documenting the status of nonconforming, inoperative, or malfunctioning SSCs and identifying the item to prevent inadvertent use.

15. Nonconforming Materials, Parts, or Components

The SAR should define the applicant's proposed provisions to control the use or disposition of nonconforming materials, parts, or components. Acceptability of the proposed provisions is contingent upon the following criteria:

a. The applicant should establish procedures to control the identification, documentation, tracking, segregation, review, disposition, and notification of affected organizations regarding nonconforming materials, parts, components, services, or activities.

b. The applicant should have adequate documentation to identify nonconforming items and describe the nonconformance, its disposition, and the related inspection requirements. The documentation should also include signature approval of the disposition.

c. The applicant should establish provisions to identify those individuals or groups with the responsibility and authority for the disposition and closeout of nonconformances.

d. Nonconforming items should be segregated from acceptable items and identified as discrepant until properly dispositioned and closed out.

e. The applicant should verify the acceptability of reworked or repaired materials, parts, and SSCs by reinspecting and retesting the item as originally inspected and tested or by a method that is at least equal to the original inspection and testing method. In addition, the applicant should document the relevant inspection, testing, rework, and repair procedures.

f. Nonconformance reports designated "accept as is" or "repair" should be made part of the inspection records and forwarded with the hardware to the customer for review and assessment.

g. The applicant should periodically analyze nonconformance reports to show quality trends and to help identify root causes of nonconformances. Significant results should be reported to responsible management for review and assessment.

16. Corrective Action

The SAR should define the applicant's proposed provisions to ensure that conditions adverse to quality are promptly identified and corrected and that measures are taken to preclude recurrence. Acceptability of the proposed provisions is contingent upon the following criteria:

a. The applicant should evaluate conditions adverse to quality (such as nonconformances, failures, malfunctions, deficiencies, deviations, and defective material and equipment) in accordance with established procedures to assess the need for corrective action.

b. The applicant should initiate corrective action to preclude recurrence of a condition identified as adverse to quality.

c. The applicant should conduct followup activities to verify proper implementation of corrective actions and to close out the corrective action documentation in a timely manner.

d. The applicant should document significant conditions adverse to quality, as well as the root causes of the conditions, and the corrective actions taken to remedy the and preclude recurrence of the conditions. In addition, this information should be reported to cognizant levels of management for review and assessment.

17. Quality Assurance Records

The SAR should define the applicant's proposed provisions for identifying, retaining, retrieving, and maintaining records that document evidence of the control of quality for activities and SSCs important to safety. Acceptability of the proposed provisions is contingent upon the following criteria:

a. The applicant should define the scope of the records program such that sufficient records will be maintained to provide documentary evidence of the quality of items and activities affecting quality. To minimize the retention of unnecessary records, the records program should list records to be retained by "type of data," rather than by record title.

b. QA records should include operating logs; results of reviews, inspections, tests, audits, and material analyses; monitoring of work performance; qualification of personnel, procedures, and equipment; and other documentation such as drawings, specifications, procurement documents, calibration procedures and reports; design review and peer review reports; nonconformance reports; and corrective action reports.

c. Records should be identified and retrievable.

d. Requirements and responsibilities for record creation, transmittal, retention (such as duration, location, fire protection, and assigned responsibilities), and maintenance subsequent to completion of work should be consistent with applicable codes, standards, and procurement documents.

e. Inspection and test records should contain the following information, where applicable:

   * a description of the type of observation
   * the date and results of the inspection or test
   * information related to conditions adverse to quality
   * identification of the inspector or data recorder
   * evidence as to the acceptability of the results
   * action taken to resolve any noted discrepancies

f. Record storage facilities should be constructed, located, and secured to prevent destruction of the records by fire, flood, theft, and deterioration by environmental conditions (such as temperature or humidity). In addition, the facilities are to be maintained by, or under the control of, the licensee throughout the life of the DCSS or the individual product.

18. Audits

The SAR should define the applicant's proposed provisions for planning and scheduling audits to verify compliance with all aspects of the QA program, and to determine the effectiveness of the overall program. Acceptability of the proposed provisions is contingent upon the following criteria:

a. The applicant should perform audits in accordance with written procedures or checklists; qualified personnel tasked with performing these audits should not have direct responsibility for the achievement of quality in the areas being audited.

b. Audit results should be documented and reviewed with management having responsibility in the area audited.

c. The applicant should establish provisions for responsible management to undertake appropriate corrective action as a followup to audit reports. Auditing organizations should schedule and conduct appropriate followup to ensure that the corrective action is effectively accomplished.

d. The applicant should perform both technical and QA programmatic audits to achieve the following objectives:

    ● Provide a comprehensive independent verification and evaluation of procedures and activities affecting quality.

    ● Verify and evaluate suppliers' QA programs, procedures, and activities.

e. Audits should be led by appropriately qualified and certified audit personnel from the QA organization. The audit team membership should include personnel (not necessarily QA organization personnel) having technical expertise in the areas being audited.

f. The applicant should schedule regular audits on the basis of the status and importance to safety of the activities being audited; such audits should be initiated early enough to ensure effective QA during design, procurement, and contracting activities.

g. The applicant should analyze and trend audit deficiency data. Resultant reports, indicating quality trends and the effectiveness of the QA program, should be given to management for review, assessment, corrective action, and followup.

h. Audits should objectively assess the effectiveness and proper implementation of the QA program and should address the technical adequacy of the activities being conducted.

i. The applicant should establish provisions requiring the performance of audits in all areas to which the requirements of the QA program apply.

APPENDIX B
MATERIALS AND COMPONENTS OF HLW STORAGE SYSTEMS

Function	Component	Safety Class*	Codes, Standards	Material	Strength (Ksi)	Surface Finish/Coat.	Contact Mat. [if Dissimilar]	Residuals on Surface	Welding Process	Weld Filler Metal
Containment	Drain port plugs									
	Inner cask bottom head									
	Inner cask lid									
	Inner cask lid seals									
	Inner cask shell									
	Inner cask shell upper hd.									
	Leak check port plug									
	Lid closure hardware									
	Outer cask inner shell									
	Outer cask lid									
	Outer cask lid seals									
	Outer cask shell bot. hd.									
	Outer cask shell upper hd.									
	Pressure relief device									
	Vent, drain, press. seals									
	Vent port plug									
Criticality Control	Basket assembly									
	Neutron absorbers									
Shielding	Gamma shielding									
	Inner cask top shield plug									
	Neutron shielding									
Heat transfer	Temp. contl. components									
Structural Integrity	Cask hardware									
	Conc. base unit, roof slab									
	Conc. struct. access bolts									
	Concrete support pad									
	Inner cask support struct.									
	Outer cask outer shell									
Opers. supp.	Access door lifting lugs									
	Lift.lugs/trunnions/grapples									
	Lift. lug/trunnion bolts									
	Protective cover									
	Roof slab lifting eyes									
	Security lockwire, seals									
	Shielded access door									
	Shielding shell									
	Transfer cask									

- The safety classification for the various functions and components should be based on terminology of NUREG/CR-6407, "Classification of Transportation Packaging and Dry Spent Fuel Storage System Components According to Importance to Safety," February, 1996. INEL Report 95/0551

APPENDIX B
MATERIALS AND COMPONENTS OF HLW STORAGE SYSTEMS
(Cont.)

Function	Component	Stress		Temp. [Storage]			Temp. [Loading]	Temp. [Unloading]	Pressure		
		Norm. (Ksi)	Max. (Ksi)	Norm. (F)	Min. (F)	Max. (F)	Max. (F/hr)	Max. (F/hr)	Min. (psig)	Max. (psig)	Gas (type)
Containment	Drain port plugs										
	Inner cask bottom head										
	Inner cask lid										
	Inner cask lid seals										
	Inner cask shell										
	Inner cask shell upper hd.										
	Leak check port plug										
	Lid closure hardware										
	Outer cask inner shell										
	Outer cask lid										
	Outer cask lid seals										
	Outer cask shell bot. hd.										
	Outer cask shell upper hd.										
	Pressure relief device										
	Vent, drain, press. seals										
	Vent port plug										
Criticality Control	Basket assembly										
	Neutron absorbers										
Shielding	Gamma shielding										
	Inner cask top shield plug										
	Neutron shielding										
Heat transfer	Temp. contl. components										
Structural Integrity	Cask hardware										
	Conc. base unit, roof slab										
	Conc. struct. access bolts										
	Concrete support pad										
	Inner cask support struct.										
	Outer cask outer shell										
Opers. supp.	Access door lifting lugs										
	Lift.lugs/trunnions/grapples										
	Lift. lug/trunnion bolts										
	Protective cover										
	Roof slab lifting eyes										
	Security lockwire, seals										
	Shielded access door										
	Shielding shell										
	Transfer cask										

APPENDIX C
GLOSSARY

ACI.   American Concrete Institute

AISC.   American Institute of Steel Constructions

ALARA.   as low as is reasonably achievable (radiation exposure)

ALI.   annual limit on intake

ANS.   American Nuclear Society

ANSI.   American National Standards Institute

API.   American Petroleum Institute

ASCE.   American Society of Civil Engineers

ASME.   American Society of Mechanical Engineers

ASTM.   American Society for Testing and Materials

AWS.   American Welding Society

B&PV.   Boiler and Pressure Vessel (ASME Code)

BWR.   boiling-water reactor

°C.   degrees Celsius

CAM.   continuous air monitor

Cask.   Dry storage system

CERCLA.   Comprehensive Environmental Response, Compensation and Liability Act (includes "Superfund")

CFR.   *U.S. Code of Federal Regulations*, organized by titles (e.g., Title 10, "Energy"), chapters (e.g., Chapter 1, "U.S. Nuclear Regulatory Commission"), parts (e.g., Part 72, "Licensing Requirements for the Independent Storage of Spent Nuclear Fuel and High-Level Radioactive Waste,"), subparts (e.g., Subpart G, "Quality Assurance"), and sections (e.g., §72.230).

Ci.   Curie

CSRP.   Cask Standard Review Plan (NUREG-1536)

CWA.   Clean Water Act

D&D.   decontamination and decommissioning

DBE   design-basis earthquake *[or "DE"]* for an ISFSI or MRS (equivalent to a "safe shutdown earthquake" for a reactor facility)

DCSS.   dry cask storage system

EA.   environmental assessment

EIS.	environmental impact statement
ER.	environmental report
°F.	degrees Fahrenheit
FONSI.	Finding of no significant impact
g.	gravitational unit (1 g = the force vertically exerted on the mass by gravity)
HTGR.	high-temperature gas-cooled reactor
HVAC.	heating, ventilation, air conditioning
I&C	Instrumentation and Control
ICRP.	International Commission on Radiological Protection
ISFSI.	independent spent fuel storage installation, which may be operated by public or private utilities, commercial entities, or governmental agencies (including the U.S. Department of Energy, DOE).
$k_{eff}.$	"k" effective. The ration of the total number of neutrons produced during a time interval (excluding neutrons produced by sources whose strengths are not a function of fission rate) to the total number of neutrons lost by absorption and leakage during the same interval.
kgf.	kilogram-force
LWR.	light-water reactor
MRS.	monitored retrievable storage facility, defined as "a complex designed, constructed, and operated by the U.S. Department of Energy for the receipt, transfer, handling, packaging, possession, safeguarding, and storage of spent nuclear fuel aged for at least 1 year, and solidified high-level radioactive waste resulting from civilian nuclear activities, pending shipment to a high-level waste repository or other disposal." [Note: DOE may also operate an ISFSI]
MSL.	mean sea level
MT.	magnetic particle test
N-stamp/ symbol.	ASME Code N-symbol (need not be applied to ISFSI or MRS components).
NA.	not applicable
NCRP.	National Council on Radiation Protection
NDT.	nil ductility temperature
NEPA.	National Environmental Policy Act
NMSS.	NRC Office of Nuclear Material Safety and Safeguards
NR.	not required
NRC.	U.S. Nuclear Regulatory Commission

NTIS.	National Technical Information Service (5285 Port Royal Road, Springfield, VA 22161), the source for NRC and other government and government-sponsored documents (including NUREG and NUREG/CR reports, Regulatory Guides, and Standard Review Plans)
NWPA.	Nuclear Waste Policy Act
ORNL	Oak Ridge National Laboratory
OSHA.	Occupational Safety and Health Administration
PT.	dye penetrant test
PWR.	pressurized-water reactor
PNNL.	Pacific Northwest National Laboratory
RAD.	unit of radiation absorbed dose
RC.	reinforced concrete
RCRA.	Resource Conservation and Recovery Act
RG.	regulatory guide (NRC)
RT.	radiographic test
Rule.	Unless used generically, a requirement stated in the *U.S. Code of Federal Regulations*.
SAR.	safety analysis report [See the related definition]
SER.	safety evaluation report [See the related definition]
SFPO.	Spent Fuel Project Office (NRC NMSS)
SNM.	special nuclear material(s)
SRP.	standard review plan
SSC.	structure, system, and/or components
SSE.	safe shutdown earthquake [derived per Appendix A to 10 CFR 100] (Used for reactor facilities; equivalent to a "Design-Basis Earthquake" or "DBE" for an ISFSI or MRS.)
std cc/s	standard cubic centimeter per second
UBC.	Uniform Building Code, published by the International Conference of Building Officials
USQ.	unreviewed safety question [used as defined by 10 CFR 72.48 for the purposes of the FSRP].
UT.	ultrasonic test
VT.	visual test
WPS.	welding procedure specification

NRC FORM 335
(2-89)
NRCM 1102,
3201, 3202

U.S. NUCLEAR REGULATORY COMMISSION

BIBLIOGRAPHIC DATA SHEET

(See instructions on the reverse)

1. REPORT NUMBER
(Assigned by NRC, Add Vol., Supp., Rev., and Addendum Numbers, if any.)

NUREG-1536

2. TITLE AND SUBTITLE

Standard Review Plan for Dry Spent Fuel Storage Systems

Final Report

3.	DATE REPORT PUBLISHED	
	MONTH	YEAR
	January	1997

4. FIN OR GRANT NUMBER

5. AUTHOR(S)

6. TYPE OF REPORT

Technical

7. PERIOD COVERED *(Inclusive Dates)*

8. PERFORMING ORGANIZATION - NAME AND ADDRESS *(If NRC, provide Division, Office or Region, U.S. Nuclear Regulatory Commission, and mailing address; if contractor, provide name and mailing address.)*

Spent Fuel Project Office
Office of Nuclear Material Safety and Safeguards
U.S. Nuclear Regulatory Commission
Washington, DC 20555-0001

9. SPONSORING ORGANIZATION - NAME AND ADDRESS *(If NRC, type "Same as above"; if contractor, provide NRC Division, Office or Region, U.S. Nuclear Regulatory Commission, and mailing address.)*

Same as 8. above

10. SUPPLEMENTARY NOTES

11. ABSTRACT *(200 words or less)*

The Standard Review Plan (SRP) for Dry Cask Storage Systems provides guidance to the Nuclear Regulatory Commission staff in the Spent Fuel Project Office for performing safety reviews of dry cask storage systems. The SRP is intended to ensure the quality and uniformity of the staff reviews and present a basis for the review scope and requirements. Part 72, Subpart B generally specifies the information needed in a license application for the independent storage of spent nuclear fuel and high level radioactive waste. Regulatory Guide 3.61 "Standard Format and Content for a Topical Safety Analysis Report for a Spent Fuel Dry Storage Cask" contains an outline of the specific information required by the staff. The SRP is divided into 14 sections which reflect the standard application format. Regulatory requirements, staff position, industry codes and standards, acceptance criteria, and other information are discussed. Comments, errors or omissions, and suggestions for improvement should be sent to the Director, Spent Fuel Project Office, U.S. Nuclear Regulatory Commission, DC 20555-0001.

12. KEY WORDS/DESCRIPTORS *(List words or phrases that will assist researchers in locating the report.)*

ISFSI, Spent Fuel, Storage Cask

13. AVAILABILITY STATEMENT

unlimited

14. SECURITY CLASSIFICATION

(This Page)

unclassified

(This Report)

unclassified

15. NUMBER OF PAGES

16. PRICE